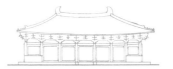

图像中国建筑史

梁思成 著

梁从诚 译

A PICTORIAL HISTORY
OF CHINESE ARCHITECTURE

A STUDY OF THE DEVELOPMENT OF ITS STRUCTURAL SYSTEM
AND THE EVOLUTION OF ITS TYPES

生活·讀書·新知 三联书店

图书在版编目（CIP）数据

图像中国建筑史 / 梁思成著；梁从诫译. —北京：
生活·读书·新知三联书店，2023.10 （2024.9 重印）
（梁思成作品）
ISBN 978-7-108-07659-5

Ⅰ.①图… Ⅱ.①梁… ②梁… Ⅲ.①建筑史－中国－
古代－图集 Ⅳ.① TU-092

中国国家版本馆 CIP 数据核字 (2023) 第 090966 号

责任编辑　刘蓉林
装帧设计　薛　宇
责任校对　张　睿
责任印制　董　欢
出版发行　生活·讀書·新知 三联书店
　　　　　（北京市东城区美术馆东街 22 号 100010）
网　　址　www.sdxjpc.com
经　　销　新华书店
印　　刷　天津裕同印刷有限公司
版　　次　2023 年 10 月北京第 1 版
　　　　　2024 年 9 月北京第 2 次印刷
开　　本　720 毫米 ×1020 毫米　1/16　印张 21.5
字　　数　130 千字　图 231 幅
印　　数　10,001－15,000 册
定　　价　108.00 元
（印装查询：01064002715；邮购查询：01084010542）

梁思成、林徽因、费慰梅（1933年摄于北平）

纪念梁思成、林徽因和他们在中国营造学社的同事们。经过他们在1931—1946年那些多灾多难的岁月中坚持不懈的努力，发现了一系列珍贵的中国古建筑遗构，并开创了以科学方法研究中国建筑史的事业。

目

录

导　言　傅熹年

译　叙　梁从诫

英文版序言　吴良镛

英文版致谢　费慰梅

英文版编辑方法　费慰梅

前　言 · · · 1

中国建筑的结构体系

起源 · · · 7

两部文法书 · · · 13

佛教传入以前和石窟中所见的木构架建筑之佐证

间接资料中的佐证 · · · 25

汉代的佐证 · · · 27

石窟中的佐证 · · · 35

木构建筑重要遗例

豪劲时期（约公元850—1050年）··· 46

醇和时期（约公元1000—1400年）··· 76

羁直时期（约公元1400—1912年）··· 109

佛　塔

古拙时期（约公元500—900年）··· 133

繁丽时期（约公元1000—1300年）··· 149

杂变时期（约公元1280—1912年）··· 165

其他砖石建筑

陵墓 ··· 180

券顶建筑 ··· 184

桥 ··· 187

台 ··· 194

牌楼 ··· 196

梁思成传略　费慰梅

附录:《图像中国建筑史》英文版全文

导　言

傅熹年

梁思成先生的英文著作《图像中国建筑史》完成于抗日战争时期，由于种种原因，他未能看到这部著作的出版。在他逝世十多年后，他的好友费慰梅女士倾力寻找，终于实现书稿的图文合璧，再经过她精心编辑，1984年美国麻省理工学院出版社将其印刷出版，1991年中国建筑工业出版社又印出汉英双语版，使这部具有极高学术价值的著作未被湮没，成为研究中国古代建筑极为重要的基础性文献。

这部著作呈现了梁思成先生在20世纪三四十年代研究中国建筑史所取得的重要学术成果，集中体现在中国建筑结构体系的发展及其形制演变方面。中国古代文化，包括建筑，延续发展数千年，从未中断，成为世界上最长寿的文化。在长期发展中，中国古代建筑逐步形成一个具有鲜明而稳定特征的独立建筑体系。20世纪初期以来，最先对中国传统建筑感兴趣的是一些外国学者。庚子之后，日本人对北京宫殿和各地著名建筑做过一些调查研究，取得了一定成果，欧洲学者也做过一些工作。20世纪30年代初，梁思成先生、刘敦桢先生加入中国营造学社，由于他们卓越的学术水平和辛勤的工作，中国的建筑史研究很快有了飞跃发展。梁先生采取由近及远、理论与实际相结合的方法，首先以清代的《工部工程做法》为课本，拜老工匠为师，以北京故宫和一些重要的明清官式建筑为标本，刻苦钻研，很快掌握了清代官式建筑的设计规律，其成果表现为《清式营造则例》一书。这之后，他又以更大的毅力钻研宋式建筑，以艰涩难懂的宋代《营造法式》为课本，实测若干宋、辽、金、元建筑，取得精确测图和数据，再与《营造法式》相印证，终于基本厘清了宋式建筑以材分°为模数的设计方法和其中所蕴含的科学成就。

梁思成先生1932年发表的《蓟县独乐寺观音阁山门考》[1]就反映了他通过精密测绘并与《营造法式》相印证，初步探明宋式建筑设计规律的过程和科学的研究方法，是这方面开天辟地的第一篇重要论文。这篇论文不仅一举超过了当时西方和日本学者研究中国建筑的水平，而且就透过形式深入探讨古代建筑设计规律而言，也超过了日本学者当时对日本建筑研究的深度。1937年，梁先生又发现了五台山佛光寺唐代大殿，这是经日本人屡次调查，印在书本上，却不识其为唐代遗构的重要古建筑。从《蓟县独乐寺观音阁山门考》到《记五台山佛光寺建筑》[2]，梁先生通过一系列对古建遗物的调查研究，把对中国古代建筑的研究方法发展得更为完备、严密，不仅科学地逐个分析了每个建筑的特点、成就，而且通过比较，掌握了各个时代的共性和不同时代的差异。正是在这个基础之上，在抗日战争最艰苦的条件下，梁先生进行了建筑通史和《营造法式》的专门研究，并以西方读者为对象，以图文并茂的方式，完成了《图像中国建筑史》的写作。

梁思成先生研究中国建筑史不仅限于技术层面，他认为"一国一族之建筑适反鉴其物质精神，继往开来之面貌"，"建筑活动与民族文化之动向实相牵连"，"中国建筑之个性乃即我民族之性格，即我艺术及思想特殊之一部，非但在其结构本身之材质方法而已"。所以，他在谈中国建筑之主要特征时，分为"结构取法及发展"和"环境思想"两大方面，把物质与技术因素和精神与社会因素相提并论。正是基于这种观点，他倾毕生精力研究建筑史，对有历史、文化和科学价值的古代城市和建筑，他主张保护；对新的建筑，他主张有中国特色，即"中而新"。他研究古代建筑不是

〔1〕参见《中国古建筑调查报告》(增补版)(上册)，生活·读书·新知三联书店即出。——编者注
〔2〕参见《中国古建筑调查报告》(增补版)(下册)，生活·读书·新知三联书店即出。——编者注

发思古之幽情，而是学以致用，却导致了其晚年的困惑。

我是 1956 年进入梁思成先生主持的清华大学建筑系与中国科学院土木建筑研究所合办的建筑历史理论研究室工作的。初到梁先生家去报到时，他同我作了很长的谈话，说中国古代建筑是延续数千年的独立体系，创造出独特的建筑风格和相应的规划设计方法。大到城市，小到单体建筑，都在世界上独树一帜，取得很高的成就，需要认真地研究总结。它既是珍贵的文化遗产，也可为今后建筑借鉴。目前有条件开展空前广泛的普查和分类分项的调查研究，在已积累了很多资料的今天，比较异同、探索规律的工作也应提到日程上来，二者相辅相成，最后才能形成有史实、有理论、能总结出发展规律的建筑史著作。当时，我是初入门的学生，对他所讲并不很理解，只能谨记，但这对我以后把研究方向转入对建筑规制和规划设计方法的探索上来，有着重要的指导作用。

费慰梅女士为《图像中国建筑史》的编辑出版倾注了大量心血。这部著作问世之后，她托友人将首批书带到广州，分寄国内友人。我至今还保留着她就此书的出版给我写来的信件。她的工作非常细致，还专门在信中对书的装帧设计加以说明，说书的封套之所以没有使用彩色照片，是因为出版社认为梁先生本人的插图都是黑白的，所以封面应该与之保持一致。她还托我向罗哲文、祁英涛、孙增蕃、王世襄等学者转达对他们为这部书的编辑出版所提供帮助的谢意。

梁思成先生以强烈的爱国热忱和严谨的科学研究，给我们留下了宝贵的学术遗产。《图像中国建筑史》见证了梁先生奋力开拓中国建筑史这门学科的努力，我们要认真学习他的著作，不断发扬光大，推动中国建筑史研究不断进步。

2022 年 9 月 15 日

译　叙[1]

梁从诫

　　先父梁思成四十多年前所著的这部书，经过父母生前挚友费慰梅女士多年的努力，历尽周折，1984 年终于在美国出版了。出版后，受到各方面的重视和好评。对于想了解中国古代建筑的西方读者来说，由中国专家直接用英文写成的这样一部书，当是一种难得的入门读物。然而，要想深入研究，只通过英文显然是不够的，在此意义上，本书的一个英汉对照本或有其特殊价值。

　　正如作者和编者所曾反复说明的，本书远非一部完备的中国建筑史。今天看来，书中各章不仅详略不够平衡，而且如少数民族建筑、民居建筑、园林建筑等等都未能述及。但若考虑到它是在怎样一种历史条件下写成的，也就难以苛求于前人了。

　　作为针对西方一般读者的普及性读物，原书使用的是一种隔行易懂的非专业性语言。善于深入浅出地解释复杂的古代中国建筑技术，是先父在学术工作中的一个特色。为了保持这一特色，译文中也有意避免过多地使用专门术语，而尽量按原文直译，再附上术语，或将后者在方括号内注出［圆括号则是英文本中原有的］。为了方便中文读者，还在方括号内作了少量其他注释。英文原书有极个别地方与作者原手稿略有出入，还有些资料，近年来已有新的研究发现，这些在译校中都已作了订

---

[1] 本文为梁从诫先生为中国建筑工业出版社 1991 年出版的汉英双语版《图像中国建筑史》所
　　写。——编者注

正或说明，并以方括号标出。在后面的英文原作上，用边码标出了各段文字在正文中相应的页码，以便使两个文本能够互相呼应。

先父的学术著作，一向写得潇洒活泼，妙趣横生，有其独特的文风。可惜这篇译文远未能体现出这种特色。父母当年曾望子成"匠"，因为仰慕宋《营造法式》修撰者将作监李诫的业绩，命我"从诫"。不料我竟然没有考取建筑系，使他们非常失望。今天，我能有机会作为隔行勉力将本书译出，为普及中国建筑史的知识尽一份微薄的力量，父母地下有知，或许会多少感到一点安慰。

先父在母亲和莫宗江先生的协助下撰写本书的时候，正值抗日战争后期，我们全家困居于四川偏远江村，过着宿不蔽风雨、食只见菜粝的生活。他们虽尝尽贫病交加、故人寥落之苦，却仍然孜孜不倦于学术研究，陋室青灯，发奋著述。那种情景，是我童年回忆中最难忘的一页。四十年后，我译此书，也可算是对于他们当时那种艰难的生活和坚毅的精神的一种纪念吧！遗憾的是，我虽忝为"班门"之后，却愧无"弄斧"之功，译文中错误失当之处一定很多，尚请父辈学者、本行专家不吝指正。

继母林洙，十年浩劫中忠实地陪伴父亲度过了他生活中最后的，也是最悲惨的一段历程，这些年来，又为整理出版他的遗著备尝辛劳。这次正是她鼓励我翻译本书，并为我核阅译文，还和清华大学建筑系资料室的同志一道为这个汉英双语版重新提供了全套原始图片供制版之用，我对她的感激是很深的。同时我也要对建筑系资料室的有关同志表示感谢。

费慰梅女士一向对本书的汉译和在中国出版一事十分关心，几年来多次来信询问我的工作进展情况。1986年初冬，我于美国与费氏二老在他们的坎布里奇家中再次相聚。四十年前，先父就是在这栋邻近哈佛大学校园的古老小楼中把本书原稿和图纸、照片托付给费夫人的。他们和我一道，又一次深情地回忆了这段往事。半个多世纪以来，他们对先父母始终不渝的友谊和对中国文化事业的积极关注，不能不使我感动。

本书译出后，曾由出版社聘请孙增蕃先生仔细校阅，在校阅过程中，又得到陈明达先生的具体指导和帮助，解决了一些专业术语的译法问题，使译文质量得以大大提高。在此谨向他们二位表示我的衷心感谢。

中国建筑工业出版社为了推动中国古建筑学术研究的发展和中外学术交流，不计经济效益方面可能受到的损失，决定以最佳印制质量出版这本书，这种精神在当今出版界已不多见，这些都令我感佩。

1991年是父亲的九十诞辰，本书能够在今年出版，更有其纪念意义。

这里我还须说明，尽管这本书基本上是根据作者原稿译出的，图版也都经重新制作，但我们还是参照了美国麻省理工学院出版社的版式和部分附录。为此，我谨向这家美国出版社致意。

最后，我还要向打字员张继莲女士致谢。这份译稿几经修改，最后得以誊清完成，与她耐心、细致的劳动也是不可分的。

1987 年 2 月于北京

1991 年 5 月补正

英文版序言[1]

吴良镛

　　杰出的中国建筑学家梁思成，是中国古建筑史研究的奠基人之一。他的这部著作［系用英文］撰写于第二次世界大战期间，当时，他刚刚完成了在华北及内地其他若干处的实地调查。梁思成教授本来计划将此书作为他的《中国艺术史》这部巨著的一部分；另一部分是中国雕塑史，他已写好了大纲。但这个计划始终未能实现。

　　现在的这部书，是他早年研究工作的一个可贵的简要总结，它可使读者对中国古建筑的伟大宝库有一个直观的概览；并通过比较的方法，了解其"有机"结构体系及其形制的演变，以及建筑的各种组成部分的发展。对于中国建筑史的初学者来说，这是一部很好的入门教材，而对于专家来说，这部书也同样有启发意义。在研究中，梁思成从不满足于已有的理解，并善于深入浅出。特别值得指出的是，由梁思成和莫宗江教授所亲手绘制的这些精美插图，将使读者获得极大的审美享受。

　　梁思成终生从事建筑事业，有着多方面的贡献。他不仅留给我们大量以中文写成的学术论文和专著（几年内将在北京出版或重刊），而且他还是一位备受尊敬的、有影响的教育家。他曾经创建过两个建筑系——1928年辽宁省的东北大学建筑系和1946年的清华大学建筑系，后者至今仍在蓬勃发展。他桃李满天下，在中国许多领

---

〔1〕这篇序是吴良镛教授应费慰梅之请，为1984年在美国出版的本书英文原作而写的。——梁
　　从诫注

域里都有他的学生在工作。1949 年以后，梁思成又全身心地投入了中国的社会主义建设事业，在中华人民共和国国徽和后来的人民英雄纪念碑的设计工作中，被任命为负责人之一。此外，他还为北京市的城市规划和促进全国文物保护做了大量有益的工作。他逝世至今虽然已经十多年，但人们仍然怀着极大的敬意和深厚的感情纪念着他。

　　这部书终于得以按照作者生前的愿望在西方出版，应当归功于梁思成的老朋友费慰梅女士。是她，在梁思成去世之后，帮助我们追回了这些已经丢失了二十多年的珍贵图版，并仔细地将文稿和大量的图稿编辑在一起，使这部书能够以现在这样的形式问世。

英文版致谢

费慰梅（Wilma Fairbank）

为了使梁思成的这部丢失了多年的著作能够如他生前所期望的那样奉献给西方读者，许多钦慕他和中国建筑的人曾共同做出过努力。其中，首先应归功于清华大学建筑系主任吴良镛教授。1980 年，是他委托我来编辑此书并设法在美国出版。我非常高兴能重新承担起三十三年前梁思成本人托付我的这个任务。

美国麻省理工学院出版社，向以刊行高质量的建筑书籍而负盛名，蒙他们同意出版本书，使这个项目得以着手进行。然而，海天相隔，怎样编好这么一部复杂的书，却是一大难题。幸运的是，我们得到了梁思成后妻林洙女士的竭诚合作。她也是清华大学建筑系的一员，对她丈夫的工作非常有认识并深情地怀念着他。我同她于 1979 年在北京相识，随后，在 1980 年和 1982 年两年中又在那里一道工作。她利用自己的业余时间，三年中和我一起不厌其烦地做了许多诸如插图的核对、标码、标题、补缺之类的细致工作，并解答了我无数的问题。我们航信频繁，她写中文，我写英文，几年未曾间断。这里，我首先要对这位亲爱的朋友表示我的感激。

1980 年夏，本书的图稿与文稿在北京得以重新合璧。此后，我曾二访北京。这些资料奇迹般的失而复得，为我敞开了回到老朋友那里去的大门。我的老友，梁思成的妹妹梁思庄、梁思成的儿子梁从诫和他的全家，还有他们的世交金岳霖都热情地接待了我，并给了我极大的帮助。我还有幸拜访了三位老一辈的建筑师，后来又和他们通信。他们是梁思成在美国宾夕法尼亚大学求学时代的同窗，又是他的至交，即现在已经故去的杨廷宝和童寯，还有陈植。在 20 世纪 30 年代曾参加过中国营造学社实地调查的较年轻的建筑史学家中，我见到了莫宗江、陈明达、罗哲文、王世襄以及刘敦桢的儿子兼学生刘叙杰。战争时期，当营造学社避难到云南、四川这些西南省份的时候，他们都在那里。还有一些更晚一辈的人，即战后 50 年代以来梁思成在清华大学的学生们。他们在本书付印前的最后阶段曾给了我特殊的帮助，特别是奚树祥、殷一和、傅熹年和他在北京中国建筑技术发展中心的同事孙增蕃等几位。本书书末的词汇表主要依靠他们四位的帮助；傅熹年和他的同事们提供了一些新的照片；奚树祥为编者注释、绘制了示意图并提供了多方面的帮助。

伦敦的安东尼·兰伯特爵士和蒂姆·罗克在重新寻得这批曾丢失的图片的过程中起了重要作用。丹麦奥胡斯大学的爱尔瑟·格兰曾对我有过重要影响，她是欧洲首屈一指的中国建筑专家，也是一位钦慕梁思成的著作的人，她曾同我一道为促使本书出版而努力。我在开始编辑本书之前，就从她那里受到过很多的教益。

在美国，我曾得到宾夕法尼亚大学、普林斯顿大学、耶鲁大学和哈佛大学档案室的慷慨帮助。在普林斯顿大学，罗伯特·索普和梁思成过去的学生黄芸生给了我指导和鼓励。我在耶鲁大学的朋友乔纳森·斯彭斯、玛丽斯·赖特、玛丽·加德纳·尼尔以及建筑师邬劲旅始终支持我的工作，特别是后者介绍给我海伦·奇尔曼女士，她是梁思成 1947 年在耶鲁大学讲学时所用的中国建筑照片的幻灯复制片的保管者。哈佛大学是我的根据地，我经常利用哈佛燕京图书馆和福格艺术博物馆，我应向哈佛燕京图书馆馆长吴文津和后者的代理主任约翰·罗森菲尔德致以特别的谢意。日本建筑史专家威廉·科尔德雷克对我总是有求必应。这里的建筑学家们也都乐于帮助我，特别是孙保禄和戴维·汉德林两位，他们一开始就是这本书的积极鼓吹者，而罗宾·布莱索则为我的编辑工作又做了校阅和加工。我的朋友琼·希尔两次为我打印誊清。我的妹妹海伦·坎农·邦德曾给予我亲切的鼓励和许多实际帮助。

美国哲学会和全国人文学科捐赠基金会资助了我的研究工作和旅行。我的北京之行不仅富有成果，而且充满乐趣，这主要应归功于加拿大驻华使馆的阿瑟·孟席斯夫妇和约翰·希金波特姆夫妇对我的热情招待。

我的丈夫费正清（John K.Fairbank）一直待在家里，这是他唯一可以摆脱一下那个斗栱世界的地方，在我编辑此书的日子里，这个斗栱世界搅得我们全家不得安生。像往常一样，他那默默的信赖和当我需要时给予我的内行的帮助总使我感激不尽。

英文版编辑方法

费慰梅

作为本书的编者，我注意到由于其原稿的奇特经历而造成的某些自身缺陷。最基本的事实当然是：本书是由梁思成在20世纪40年代写成的，而他却在其出版前十多年便去世了。如果作者健在，很可能会将本书的某些部分重新写过或加以补充。特别引人注意的是，他在对木构建筑和塔作了较为详尽的阐述之后，却只对桥、陵墓和其他类型的建筑作了非常简略的评论。在这种情况下，我的责任在于严格忠于他的原意，尽量采用他的原文而不去擅自改动他的打字原稿。我插入的一段解释是特别标明了的。

当梁思成决定将中国营造学社1931—1937年在华北地区以及1938—1946年抗战期间在四川和云南的考察成果向西方读者作一个简明介绍的时候，曾打算采用图像历史的形式。本书的标题可能使人以为这是一部全面的建筑史，但事实上这本书仅仅简介了营造学社在华北地区和其他省份中曾经考察过的某些重要的古建筑。作者原来只准备选用学社档案中的一些照片，附以根据实测绘制的这些古建筑的平面、立面和断面图，以此来说明中国建筑结构的发展，因而只在图版中作一些英汉对照的简要注释而未另行撰文。1943年，这些图版按计划绘制完毕，梁氏将其携至重庆，由他在美国军事情报处摄影室工作的朋友们代他翻拍成两卷缩微胶片。其中一卷他交给我保管以防万一，我后来曾将其存入哈佛燕京图书馆；而当1947年他将原图带到美国时，又把另一个胶卷留在了国内。这种防范措施后来导致了有趣的结果。

与此同时，梁氏承认："在图版绘成之后，又感到几句解说可能还是必要的。"这部文稿，阐述了他对自己多年来所作的开创性的实地考察工作的分析和结论。它简明扼要，却涉及数量惊人的实例。既是摘要，就不可能把梁氏在《中国营造学社汇刊》探讨这些建筑的文章中关于中国建筑的许多方面的论点都包括进去，他为各个时期所标的名称反映了当时他对这个时期的评价。自那时迄今已近四十年。但人们期待这书已久，看来最好还是将它作为一份历史文献而保留其原貌，不作改动。原稿最后有两页介绍皇家园林的文字，因尚未完稿而且离开建筑结构的发展这一主题较远，所以删去了。

梁氏的原稿是用简单明了的英文写成的，其中重要的建筑术语都按威妥玛系统拼音，现除个别专门名词因个人的取舍而有不同之外，均照原样保留。

书后技术术语一览是为了向［西方］读者解释它们的含义，并附有汉字。这种英汉对照的形式十分重要。图版中以两种文字书写图注及说明，其含义是使那些想认真研究中国建筑的西方学生明白，他们必须熟悉那些术语、重要建筑及其所在地的中文名称。梁思成在宾大时曾学过那些西方建筑史上类似的英法对照名词，在这个基础上，他又增加了自己经过艰苦钻研而获得的关于中国建筑史的知识。同样，西方未来几代的学生们要想经常方便地到中国旅行，就需懂得那里的语言和以往被人忽视的建筑艺术。

本书所用的大部分照片反映了那些建筑物在 20 世纪 30 年代的面貌。几座现已不存的建筑都在说明中注明"已毁"，但那究竟是由于自然朽坏，还是意外事故乃至有意破坏，则不得而知。在图版中手书的图注里有若干小错误，个别英文、罗马拼音或年代不确，都有意识地不予更改。遗憾的是，1947 年梁思成由于个人原因，未及在全书之后略作提要。但他那些首次发表于此的"演变图"（图 20，图 21，图 32，图 37，图 38，图 63）已清楚地概括了他在书中所述及的那些关键性的演变。书中其他各图三十年来曾在东、西方多次发表，却从未说明其作者是梁思成或营造学社。它们的来源就是梁氏 1943 年翻拍的第二卷缩微胶片。1952 年，在一份供清华学生使用的标明"内部参考"的小册子中，曾将这些图版翻印出来，而这本小册子后来却流传到了欧洲和英国。由于小册子有图无文，所以，也许不能指责有些人剽窃。无论如何，本书的出版应当使上述做法就此打住。

前　言

这本书全然不是一部完备的中国建筑史，而仅仅是试图借助若干典型实例的照片和图解来说明中国建筑结构体系的发展及其形制的演变。最初我曾打算完全不用释文，但在图纸绘成之后，又感到几句解说可能还是必要的，因此，才补写了这篇简要的文字。

中国的建筑是一种高度"有机"的结构。它完全是中国土生土长的东西：孕育并发祥于遥远的史前时期；"发育"于汉代（约在公元开始的时候）；成熟并逞其豪劲于唐代（7—8世纪）；臻于完美醇和于宋代（11—12世纪）；然后于明代初叶（15世纪）开始显出衰老羁直之象。虽然很难说它的生命力还能保持多久，但至少在本书所述及的三十个世纪之中，这种结构始终保持着自己的机能，而这正是从这种条理清楚的木构架的巧妙构造中产生出来的；其中每个部件的规格、形状和位置都取决于结构上的需要。所以，研究中国的建筑物首先就应剖析它的构造。正因为如此，其断面图就比其立面图更为重要。这是和研究欧洲建筑大相异趣的一个方面；也许哥特式建筑另当别论，因为它的构造对其外形的制约作用比任何别种式样的欧洲建筑都要大。

如今，随着钢筋混凝土和钢架结构的出现，中国建筑正面临着一个严峻的局面。诚然，在中国古代建筑和最现代化的建筑之间有着某种基本的相似之处，但是，这两者能够结合起来吗？中国传统的建筑结构体系能够使用这些新材料并找到一种新的表现形式吗？可能性是有的。但这绝不应是盲目地"仿古"，而必须有所创新。否则，中国式的建筑今后将不复存在。

对中国建筑进行全面研究，就必须涉及日本建筑。因为按正确的分类来说，某些早期的日本建筑应被认为是自中国传入的。但是，关于这个问题，在这本简要的著作中只能约略地提到。

请读者不要因为本书所举的各种实例中绝大多数是佛教的庙宇、塔和墓而感到意外。须知，不论何时何地，宗教都曾是建筑创作的一个最强大的推动力量。

本书所用资料，几乎全部选自中国营造学社的学术档案，其中一些曾发表于《中国营造学社汇刊》。这个研究机构自1929年创建以来，在社长朱启钤先生和战争年代（1937—1946）中的代理社长周贻春博士的富于启发性的指导之下，始终致力于在全国系统地寻找古建筑实例，并从考古与地理学两个方面对它们加以研究。到目前为止，已对十五个省内的两百余县进行了调查，若不是战争的干扰使实地调查几乎完全停顿，我们肯定还会搜集到更多的实例。而且，当我此刻在四川省西部这个偏僻的小村中撰写本书时，由于许多资料不在手头，也使工作颇受阻碍。当营造学社迁往内地时，这些资料被留在北平。同时，书中所提及的若干实例，肯定已毁于战火。它们遭到破坏的程度，只有待对这些建筑物逐一重新调查时才能知道。

营造学社的资料，是在多次实地调查中收集而得的。这些实地调查，都是由原营造学社文献部主任，现中央大学工学院院长、建筑系主任刘敦桢教授或我本人主持的。蒙他惠允我在书中引用他的某些资料，谨在此表示深切的谢意。我也要对我的同事、营造学社副研究员莫宗江先生致谢，我的各次实地考察几乎都有他同行；他还为本书绘制了大部分图版。

我也要感谢中央研究院历史语言研究所考古部主任李济博士和该所副研究员石璋如先生，承他们允许我复制了安阳出土的殷墟平面图；同时对于作为中央博物院院长的李济博士，我还要感谢他允许我使用中国营造学社也曾参加的江口汉墓发掘中的某些材料。

我还要感谢我的朋友和同事费慰梅女士［费正清夫人］。她是中国营造学社的成员，曾在中国做过广泛旅行，并参加过我的一次实地调查活动。我不仅要感谢她所做的武梁祠和朱鲔墓石室的复原工作，而且要感谢她对我的大力支持和鼓励，因而使得本书的编写工作能够大大加快。我也要感谢她在任驻重庆美国大使馆文化参赞期间，百忙中抽时间耐心审读我的原稿，改正我英文上的错误。她在上述职务中为加强中美两国之间的文化交流做出了极有价值的贡献。

最后，我要感谢我的妻子、同事和旧日的同窗林徽因。二十多年来，她在我们共同的事业中不懈地贡献着力量。从在大学建筑系求学的时代起，我们就互相为对方"干苦力活"，以后，在大部分的实地调查中，她又与我做伴，有过许多重要的发现，并对众多的建筑物进行过实测和草绘。近年来，她虽罹患重病，却仍葆其天赋的机敏与坚毅；在战争时期的艰难日子里，营造学社的学术精神和士气得以维持，主要应归功于她。没有她的合作与启迪，无论是本书的撰写，还是我对中国建筑的任何一项研究工作，都是不可能成功的。[1]

〔1〕
1946年梁思成对之表示感谢的人们，多数都已故去。目前，除我本人之外，只有北京清华大学的莫宗江和台北"中央研究院"的石璋如尚在。——费慰梅注

梁思成

识于四川省李庄

中国营造学社战时社址

1946年4月

中国建筑的结构体系

起　源

中国的建筑与中国的文明同样古老。所有的资料来源——文字、图像、实例——都有力地证明了中国人一直采用着一种土生土长的构造体系，从史前时期直到当代，始终保持着自己的基本特征。在中国文化影响所及的广大地区里——从新疆到日本，从东北三省到印支半岛北方，都流行着这同一种构造体系。尽管中国曾不断地遭受外来的军事、文化和精神侵犯，这种体系竟能在如此广袤的地域和长达四千余年的时间中常存不败，且至今还在应用而不易其基本特征，这一现象，只有中华文明的延续性可以与之相提并论，因为，中国建筑本来就是这一文明的一个不可分离的组成部分。

在河南省安阳市市郊，在经中央研究院发掘的殷代帝王们（约公元前 1766—约公元前 1122 年）的宫殿和墓葬遗址中，发现了迄今所知最古老的中国房屋遗迹（图 10）。这是一些很大的黄土台基，台上以有规则的间距放置着一些未经加工的砾石，较平的一面向上，上面覆以青铜圆盘（后世称之为櫍）。在这些铜盘上，发现了已经炭化的木材，是一些木柱的下端，它们曾支承过上面的建筑。这些建筑是在周人征服殷王朝（约公元前 1122 年）并掠夺这座帝都时被焚毁的。这些柱础的布置方式证明当时就已存在着一种定型，一个伟大的民族及其文明从此注定要在其庇护之下生存，直到今天。

这种结构体系的特征包括：一个高起的台基，作为以木构梁柱为骨架的建筑物的基座，再支承一个外檐伸出的坡形屋顶（图 1、图 2）。

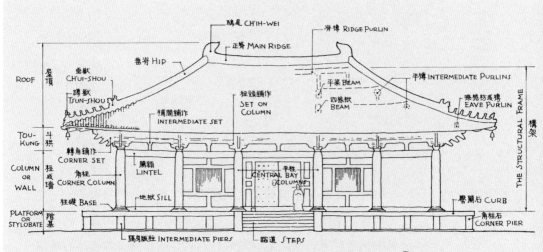

NAMES OF PRINCIPAL PARTS OF A CHINESE BUILDING

中國建築主要部份名稱圖

这种骨架式的构造使人们可以完全不受约束地筑墙和开窗。从热带的印支半岛到亚寒带的东北三省,人们只需简单地调整一下墙壁和门窗间的比例就可以在各种不同的气候下使其房屋都舒适合用。正是由于这种高度的灵活性和适应性,使这种构造方法能够适用于任何华夏文明所及之处,使其居住者能有效地躲避风雨,而不论气候有多大差异。在西方建筑中,除了英国伊丽莎白女王时代的露明木骨架建筑这一有限的例外,直到 20 世纪发明钢筋混凝土和钢框架结构之前,可能还没有与此相似的做法。

英文版编辑注释:曲面屋顶与斗栱

图 1 和图 2 表示了中国传统建筑的基本特征,其表现方式对于熟悉中国建筑的人来说是明白易懂的。然而,并不是每一位读者都有看到中国建筑实物或研究中国木构架建筑的机会,为此,我在这里再作一点简要的说明。

中国殿堂建筑最引人注目的外形,就是那外檐伸出的曲面屋

图 2   2

中国建筑之"柱   The Chinese "order"
式"（本图是梁氏   （the most frequently
所绘图版中最常被   reprinted of Liang's
人复制的一张）   drawings）

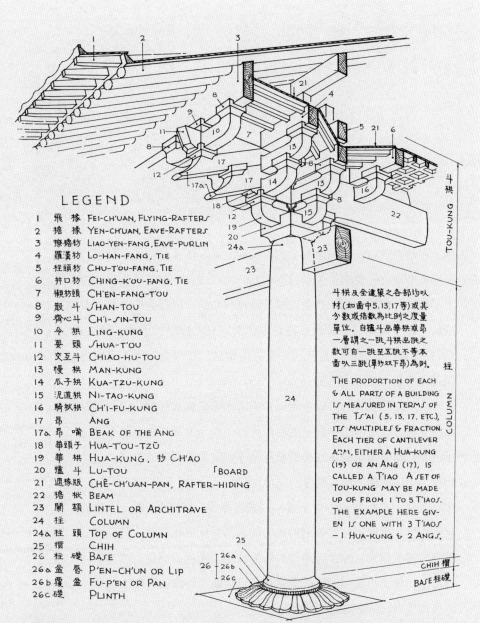

LEGEND

1 飛 椽 FEI-CH'UAN, FLYING-RAFTERS
2 檐 椽 YEN-CH'UAN, EAVE-RAFTERS
3 撩檐枋 LIAO-YEN-FANG, EAVE-PURLIN
4 羅漢枋 LO-HAN-FANG, TIE
5 柱頭枋 CHU-T'OU-FANG, TIE
6 井口枋 CHING-K'OU-FANG, TIE
7 襯枋頭 CH'EN-FANG-T'OU
8 散 斗 SHAN-TOU
9 齊心斗 CH'I-SIN-TOU
10 令 拱 LING-KUNG
11 耍 頭 SHUA-T'OU
12 交互斗 CHIAO-HU-TOU
13 慢 拱 MAN-KUNG
14 瓜子拱 KUA-TZU-KUNG
15 泥道拱 NI-TAO-KUNG
16 騎栿拱 CH'I-FU-KUNG
17 昂 ANG
17a 昂 嘴 BEAK OF THE ANG
18 華頭子 HUA-T'OU-TZŬ
19 華 拱 HUA-KUNG, 抄 CH'AO
20 櫨 斗 LU-TOU
21 遮椽版 CHÊ-CH'UAN-PAN, RAFTER-HIDING「BOARD
22 檐 栿 BEAM
23 闌 額 LINTEL OR ARCHITRAVE
24 柱 COLUMN
24a 柱 頭 TOP OF COLUMN
25 櫍 CHIH
26 柱 礎 BASE
26a 盆 唇 P'EN-CH'UN OR LIP
26b 覆 盆 FU-P'EN OR PAN
26c 礎 PLINTH

斗栱及全建築之各部均以
材（如圖中5.13.17等）或其
分數或倍數為比例之度量
單位. 自櫨斗出華栱或昂
一層謂之一跳, 斗栱出跳之
數可自一跳至五跳不等本
圖叭三跳（單抄双下昂）為叭.

THE PROPORTION OF EACH
& ALL PARTS OF A BUILDING
IS MEASURED IN TERMS OF
THE TS'AI (5. 13. 17. ETC.),
ITS MULTIPLES & FRACTION.
EACH TIER OF CANTILEVER
ARM, EITHER A HUA-KUNG
(19) OR AN ANG (17), IS
CALLED A T'IAO A SET OF
TOU-KUNG MAY BE MADE
UP OF FROM 1 TO 5 T'IAOS.
THE EXAMPLE HERE GIV-
EN IS ONE WITH 3 T'IAOS
— 1 HUA-KUNG & 2 ANGS.

CHIH 櫍
BASE 柱礎

中國建築之"ORDER". 斗栱.檐柱.柱礎   THE CHINESE "ORDER"

顶。屋顶由立在高起的阶基上的木构架支承。图3显示了五种屋顶构造类型的九种变形。关于这些形制，梁思成在本书都已列出。为了了解这些屋檐上翘的曲面屋顶是怎样构成的，以及它们为什么要造成这样，我们就必须研究这种木构架本身。按照梁思成的说法："研究中国的建筑物首先就应剖析它的构造。正因为如此，其断面图就比其立面图更为重要。"

从断面图上我们可以看到，在中国木构架建筑的构造中，对屋顶的支承方式根本上不同于通常的西方三角形屋顶桁架，而正是由于后者，西方建筑的直线形的坡屋顶才会有那样僵硬的外表。与此相反，中国的框架则有明显的灵活性（图4）。木构架由柱和梁组成。梁有几层，其长度由下而上逐层递减。平槫〔檩条〕，即支承椽的水平构件，被置于层层收缩的构架的肩部。椽都比较短，其长度只有槫与槫之间距。工匠可通过对构架高度与跨度的调整，按其所需而造出各种大小及不同弧度的屋顶。屋顶的下凹曲面可使半筒形屋瓦严密接合，从而防止雨水渗漏。

图3
屋顶的五种类型
1- 悬山；2- 硬山；3- 庑殿；4、6- 歇山；5- 攒尖；7~9- 分别为 5、4〔原文误为6〕和 3 的重檐式

3
Five types gable roof
1.overhanging gable roof; 2.flush gable roof; 3.hip roof; 4 and 6.gable-and-hip roofs; 5.pyramidal roof; 7~9.double-eaved version of 5,6 and 3 respectively

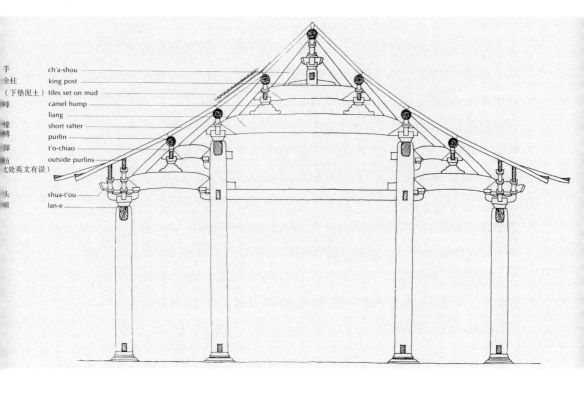

手全 (下垫泥土) 峰 椽槫栿 (七处英文有误) 头额

ch'a-shou
king post
tiles set on mud
camel hump
liang
short rafter
purlin
t'o-chiao
outside purlins

shua-t'ou
lan-e

图4
断面图，表现出灵活的梁柱框架支承着曲面屋顶

4
Section, showing flexible beam skeleton supporting curved roof

屋檐向外出跳的深度也是值得注意的。例如，由梁氏所发现的唐代佛光寺大殿（公元857年），其屋檐竟从下面的檐柱向外挑出约14英尺［4米］（图24）。能够保护这座木构建筑历经一千一百多年的风雨而不毁，这种屋檐所起的重要作用是显而易见的。例如，它可以使沿曲面屋顶瓦槽顺流而下的雨水泻向远处。

然而，屋檐上翘的直接功能还在于使房屋虽然出檐很远，但室内仍能有充足的光线。这就需要使支承出挑屋檐的结构一方面必须从内部构架向外大大延伸，另一方面又必须向上抬高以造成屋檐的翘度。这些是怎样做到的呢？

正如梁氏指出的："是斗栱（托架装置）起了主导作用。其作用是如此重要，以致如果不彻底了解它，就根本无法研究中国建筑。它构成了中国建筑'柱式'中的决定性特征。"图5是一组斗栱的等角投影图，图6则是一组置于柱头上的斗栱。这里我们又遇到了一个陌生的东西。在西方建筑中，我们习惯于那种简单的柱头，这种柱头直接承重并将荷载传递到柱上，而斗栱却是一个十分复杂的部件。虽然其底部只是柱头上的一块大方木，但从其中却向四面伸出十字形的横木［栱］。后者上面又置有较小的方木［斗］，从中再次向四面伸出更长的横木以均衡地承托更在上的部件。这种前伸的横木［华栱］以大方木块为支点一层层向上和向外延伸，即称为"出跳"，以支承向外挑出的屋檐的重量。它们在外部所受的压力，由这一托架［斗栱］内部所承受的重量来平衡。在这套托架中有与华栱交叉而与墙面平行的横向栱。从里面上部结构向下斜伸的悬臂长木称为昂，它以栌斗为支点，穿过斗栱向外伸出，以支承最外面的橑檐枋（图6），这种外面的荷载是以里面上部的槫或梁对昂尾的下压力来平衡的。向外突出的尖形昂嘴很容易在斗栱中被识别出来。梁思成在本书中对于这种构造在其演变过程中的各种复杂情况都作了较详尽的解释。

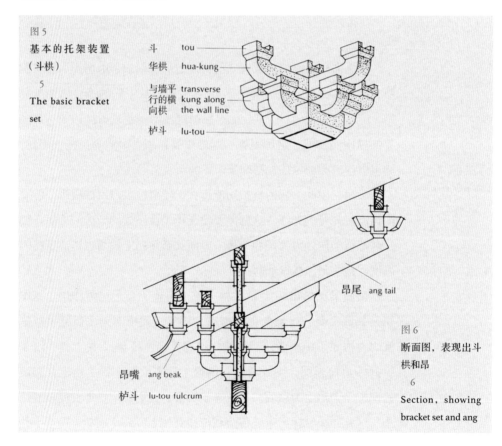

图 5
基本的托架装置（斗栱）

斗 tou
华栱 hua-kung
与墙平行的横向栱 transverse kung along the wall line
栌斗 lu-tou

5
The basic bracket set

昂尾 ang tail

昂嘴 ang beak
栌斗 lu-tou fulcrum

图 6
断面图，表现出斗栱和昂

6
Section, showing bracket set and ang

图像中国建筑史

两部文法书

随着这种体系的逐渐成熟，出现了设计和施工中必须遵循的一整套完备的规程。研究中国建筑史而不懂得这套规程，就如同研究英国文学而不懂得英文文法一样。因此，有必要对这些规程作一简略的探讨。

幸运的是，中国历史上两个曾经进行过重大建筑活动的时代有两部重要的书籍传世：宋代（公元 960—1279 年）的《营造法式》和清代（公元 1644—1912 年）的《工程做法则例》，我们可以把它们称为中国建筑的两部"文法书"。它们都是官府颁发的工程规范，因而对于研究中国建筑的技术来说，是极为重要的。今天，我们之所以能够理解各种建筑术语，并在对不同时代的建筑进行比较研究时有所依据，都因为有了这两部书。

《营造法式》

《营造法式》是宋徽宗在位时（公元 1101—1125 年）朝廷中主管营造事务的将作监李诫编撰的。全书共三十四卷，其中十三卷是关于基础、城寨、石作及雕饰，以及大木作（木构架、柱、梁、枋、额、斗栱、槫、椽等）、小木作（门、窗、扇、屏风、平棊、佛龛等）、砖瓦作（砖瓦及瓦饰的官式等级及其用法）和彩画作（即彩画的官式等级和图样）的；其余各卷是各类术语的释义及估算各种工、料的数据。全书最后四卷是各类木作、石作和彩画的图样。

本书出版于公元 1103 年［崇宁二年］。但在以后的八个半世纪中，由于建筑在术语和形式方面都发生了变化，更由于在那个时代的环境中，文人学者轻视技术和体力劳动，竟使这部著作长期湮没无闻，而仅仅被一些收藏家当作稀世奇书束诸高阁。对于今天的外行人来说，它们极其难懂，其中的许多章节和术语几乎是不可理解的。然而，经过中国营造学社同仁的悉心努力，首先通过对清代建筑规范的掌握，以后又研究了已发现的相当数量的建于 10—12 世纪木构建筑实例，书中的许多奥秘终于被揭开，从而使它现在成为一部可以读得懂的书了。

由于中国建筑的主要材料是木材，对于理解中国建筑结构体系来说，书中的"大木作"部分最为重要。其基本规范我们已在图 2 至图 7 中作了图解，并可归纳如下。

一、量材单位——材和栔

材这一术语，有两种含义：

（甲）某种标准大小的，用以制作斗栱中的栱的木料，以及所有高度和宽度都与栱相同的木料。材分为八种规格，视所建房屋的类型和官式等级而定。

（乙）一种度量单位，其释义如下：

材按其高度均分为十五，各为一分°[1]。材的宽度为十分。房屋的高度和进深，所使用的全部构件的尺寸，屋顶举折的高度及其曲线，总之，房屋的一切尺寸，都按其所用材的等级中相应的分° 为度。

当一材在使用中被置于另一材之上时，通常要在两材之间以斗垫托其空隙，其空隙距离为六分°，称为栔。材的高度相当于一材加一栔的，称为足材。在宋代，一栋房屋的规格及其各部之间的比例关系，都以其所用等级木料的材、栔、分° 来表示。[2]

二、斗栱

一组斗栱［宋代称为"朵"，清代称为"攒"］是由若干个斗

[1]
根据梁思成先生在《〈营造法式〉注释》中所作说明，为了避免混淆，凡分寸之"分"皆如字，材分之"分"音符问切，一律加符号写成"分°"。——编者注

[2]
根据后来的研究，《营造法式》在结构上所用的基本度量单位实际上是"分"，即"材"高的十五分之一。——孙增蕃校注

（方木）和栱（横材）组合而成的。其功能是将上面的水平构件的重量传递到下面的垂直构件上去。斗栱可置于柱头上，也可置于两柱之间的阑额或角柱上。根据其位置它们分别被称为"柱头铺作""补间铺作"或"转角铺作"["铺作"即斗栱的总称]。组成斗栱的构件又分为斗、栱和昂三大类。根据位置和功能的差异，共有四种斗和五种栱。然而从结构方面说，最重要的还是栌斗（即主要的斗）和华栱。后者是从栌斗向前后挑出的，与建筑物正面成直角的栱。有时华栱之上还有一个斜向构件，与地平约成 30 度交角，称为昂。它的上端，称为昂尾，常由梁或槫的重量将其下压，从而成为支承挑出的屋檐的一根杠杆。

华栱也可以上下重叠使用，层层向外或向内挑出，称为出跳。一组斗栱可含一至五跳。横向的栱与华栱在栌斗上相交叉。每一跳有一至二层横向栱的，称为"计心"；没有横向栱的出跳称为"偷心"。只有一层横向栱的，称为单栱；有两层横向栱的，称为重栱。依出跳的数目、计心或偷心的安排、华栱和昂的悬挑以及单栱和重栱的使用等不同情况，可形成斗栱的多种组合方式。

三、梁

梁的尺寸和形状因其功能和位置的不同而异。天花下面的梁栿称为明栿，即"外露的梁"；它们或为直梁，或为稍呈弓形的月梁，即"新月形的梁"。天花以上不加刨整的梁称为草栿，用以承受屋顶的重量。梁的周径依其长度而各不相同，但作为一种标准，其断面的高度与宽度总保持着三与二之比。

四、柱

柱的长度与直径没有什么严格的规定。其直径可自一材一契至三材不等。柱身通直或呈梭形，后者自柱的上部三分之一处开始依曲线收缩["卷杀"]。用柱之制中最重要的规定是：（1）柱高自当心间往两角逐渐增加["生起"]；（2）各柱都以约 1：100 的比例略

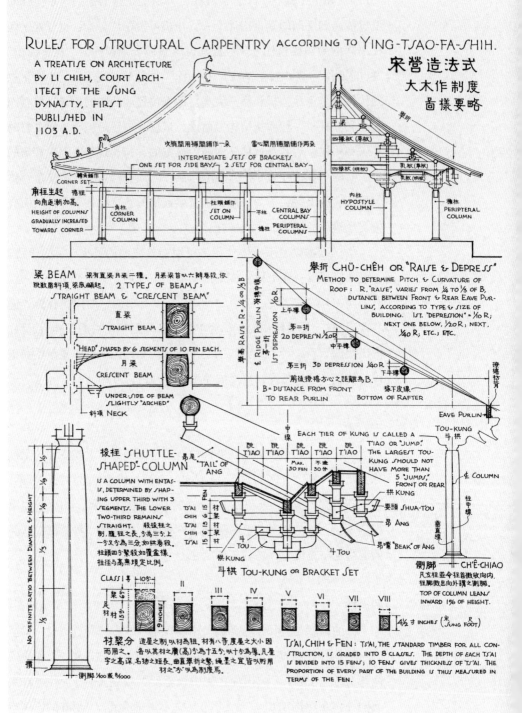

向内倾［"侧脚"］。这些手法有助于使人产生一种稳定感。

五、曲面的屋顶（举折）

屋顶横断面的曲线是由举（即脊槫［檩］的升高）和折（即椽线的下降）所造成的。其坡度决定于屋脊的升高程度，可以从一般小房子的 1:2 到大殿堂的 2:3 不等。升高的高度称为举高。屋顶的曲线是这样形成的：从脊槫到橑檐枋背之间画一直线，脊槫以下第一根槫的位置应按举高的十分之一低于此线；从这槫到橑檐枋背再画一直线，第二根槫的位置应按举高的二十分之一低于此线；依此类推，每根槫降低的高度递减一半。将这些点用直线连接起来，就形成了屋顶的曲线。这一方法称为折屋，意思是"将屋顶折弯"。

除以上这些基本规范外，《营造法式》中还分别详尽地叙述了宋代关于阑额、枋、角梁、槫、椽和其他部件的用法和做法。

仔细研究本书以后各章中关于不同时代大木作演化情况，可以使我们对中国建筑结构体系的发展历史有一个清楚的了解。

《营造法式》中小木作各章，是有关门、窗、槅扇、屏风以及其他非结构部件的设计规范。其传统做法，后世大体继承了下来而无重大改变。平槫都呈正方或长方形，藻井常饰以小斗栱。佛龛和道帐也常富于建筑特征，并以斗栱作为装饰。

在有关瓦及瓦饰的一章里，详述了依建筑规定的大小等级，屋顶上用以装饰屋脊的鸱、尾、蹲兽应取的规格和数目。虽然至今在中国南方仍然通行用板瓦叠成屋脊的做法，但在殿堂建筑中则久已不用了。

在关于彩画的各章里，列出了不同等级的房屋所应使用的各种类别的彩画；说明了其用色规则，主要是冷暖色对比的原则。从中可以了解，色彩的明暗是以不同深浅的同一色并列叠晕而不是以单色的加深来表现的。主要的用色是蓝、红和绿，缀以墨、白；有时也用黄色。用色的这种传统自唐代（公元 618—907 年）一直延续至今。

图 7
宋《营造法式》大木作制度图样要略
7
Sung dynasty
rules for structural
carpentry

《营造法式》还以不少篇幅详述了各种部件和构件的制作细节。如斗、栱和昂的斫造和卷杀；怎样使梁成为弓形并使其两侧微凸；怎样为柱基和勾栏雕刻饰纹；以及不同类型和等级的彩画的用色调配；等等。按现代的含义，《营造法式》在许多方面确是一部教科书。

《工程做法则例》

《工程做法则例》是公元 1734 年〔清雍正十二年，原图注误为 1733 年〕由工部刊行的。前二十七卷是二十七种不同建筑如大殿、城楼、住宅、仓库、凉亭等等的筑造规则。每种建筑物的每个构件都有规定的尺寸。这一点与《营造法式》不同，后者只有供设计和计算时用的一般规则和比例。次十三卷是各式斗栱的尺寸和安装法，还有七卷阐述了门、窗、槅扇、屏风，以及砖作、石作和土作的做法。最后二十四卷是用料和用工的估算。

这部书只有二十七种建筑的断面图共二十七幅。书中没有关于具有时代特征的建筑细节的说明，如栱和昂的成形方法、彩画的绘制等等。幸而大量的清代建筑实物仍在，我们可以方便地对之加以研究，从而弥补了此书之不足。

从《工程做法则例》的前四十七卷中，可以归结出若干原则来，而其中与大木作或结构设计有关的，主要是以下几项。图 8 对这几项作了图解。

一、材的高度减少

如上所述，依宋制，材高 15 分°（宽 10 分°），栔为 6 分°，故足材的高度为 21 分°。而清代，关于材、栔、分°的概念在匠师们头脑中似已不存，而作为承受栱的部位——斗口，即斗上的卯口，却成了一个度量标准。它与栱同宽，因此即相当于材宽（宋制为 10 分°）。斗栱各部分的尺寸及比例都以斗口的倍数或分数为度。上、下两栱之间仍为 6 分°（清代称为 0.6 斗口），栱的高度由 15 分°

图 8

清《工程做法则例》大式大木图样要略

8

Ch'ing dynasty rules for structural carpentry

图像中国建筑史

RULES FOR STRUCTURAL CARPENTRY ACCORDING TO KUNG-CH'ENG-TSO-FA

清工程做法則例
雍正十二年工部頒布刊行

大式大木
畫樣要略

OFFICIAL REGULATIONS FOR ARCHITECTURAL DESIGN IN THE CH'ING DYNASTY, PUBLISHED BY THE MINISTRY OF WORKS IN 1733.

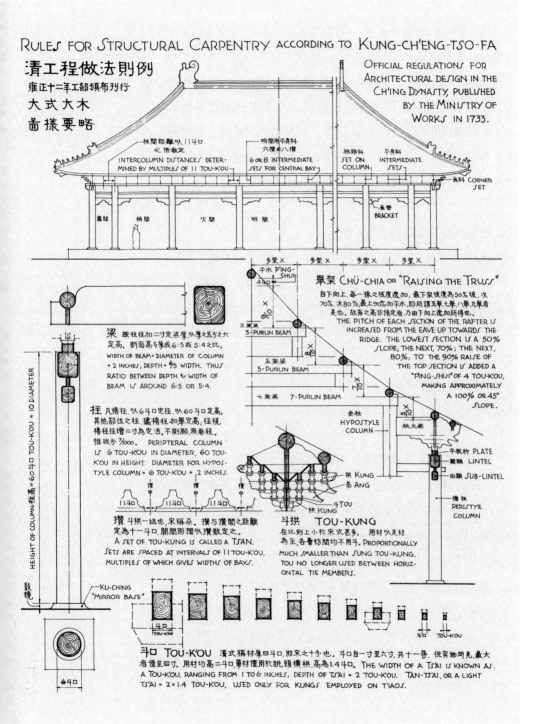

减为 14 分°（称为 1.4 斗口）。此时，足材仅高 20 分°，即 2 斗口。

宋、清两代的斗栱还有一个重要区别。沿着与建筑物正面平行的柱心线上放置的斗，称为栌斗，栌斗中交叉地放着伸出的栱，其上则支承着几层材。在宋代，这几层材之间用斗托垫，其间的空隙，或露明，或用灰泥填实。在清代，这几层材却直接叠放在一起，每层厚度相当于 2 斗口。这样，材与材之间放置斗的空隙，即栔，便被取消了。这些看来似乎是微不足道的变化使斗栱的外形大为改观，人们可一望而知。

二、柱径与柱高之间的规定比例

清代《则例》规定，柱径为 6 斗口（即宋制 4 材），高 60 斗口，即径的十倍。据《营造法式》，宋代的柱径从不超过 3 材，其高度则由设计者任定。这样，从比例上看：清代的柱比宋代加大很多，而斗栱却缩小很多，以致竟纤小得成为无关紧要的东西。其结果，两柱之间的斗栱［宋称"补间铺作"，清称"平身科"］数比过去大大增加，有时竟达七八攒之多；而在宋代，依《营造法式》的规定和所见实例，其数目从不超过两朵。

三、建筑的面阔及进深取决于斗的数目

由于两柱间斗栱攒数增加，两攒斗栱之间的距离便严格地规定为 11 斗口中到中。其结果，柱与柱的间距，进而至于全屋的面阔和进深，都必须相当于 11 斗口的若干倍数。

四、建筑物立面所有柱高都相等［"角柱不生起"］

宋代那种柱高由中央向屋角逐步增加的做法［"生起"］已不再继续。柱身虽稍呈锥形，但已成直线，而无卷杀。这样，清代建筑比之宋代，从整体上显得更僵直一些，但各柱仍遵循略向内倾［"侧脚"］的规定。

五、梁的宽度增加

在宋代，梁的高度与宽度之比大体上是 3∶2。而依清制，这一比例已改为 5∶4 或 6∶5，显然是对材料力学的无知所致；更有甚者，还规定梁宽一律为"以柱径加二寸定厚"。看来这是最武断、最不合理的规定。清代所有的梁都是直的，在其官式建筑中已不再使用月梁。

六、屋顶的坡度更陡

宋代称为"举折"的做法，在清代称为"举架"，即"举起屋架"之意。两种做法的结果虽大体相仿，但其基本概念却完全相异。宋代建筑的脊槫高度是事先定好了的，屋顶坡形曲线是靠下折以下诸槫而形成的；清代的匠师们却是由下而上，使其第一步即最低的两根槫的间距的举高为"五举"，即 5∶10 的坡度；第二步为"六举"，即 6∶10 的坡度；第三步为 6½∶10；第四步为 7½∶10；如此直到"九举"，即 9∶10 的坡度，而脊槫的位置要依各步举高的结果而定。由此形成的清代建筑屋顶的坡度，一般都比宋代要陡。这一点使人们很容易区分建筑物的年代。

下面我们将会看到，清代建筑一般的特征是：柱和过梁外形刻板、僵直；屋顶坡度过分陡峭；檐下斗栱很小。可能是《工程做法则例》中那些严格的、不容变通的规矩和尺寸，竟使《营造法式》时代的建筑那种柔和秀丽的动人面貌丧失殆尽。

这两部书都没有提到平面布局问题。《营造法式》中有几幅平面图，但不是用以表明内部空间的分割，而只是表明柱的配置。中国建筑与欧洲建筑不同，无论是庙宇还是住宅，都很少在平面设计时将独立单元的内部再行分割。由于很容易在任何两根柱子之间用槅扇或屏风分割，所以内部平面设计的问题几乎不存在，总体平面设计则涉及若干独立单位的群体组合。一般的原则是，将若干建筑物安排在一个庭院的四周，更确切地说，是通过若干建筑物的安排来形成一个庭院或天井。各建筑物之间有时以走廊相连，但较小的住宅则没有。一所大住宅常由沿着同一条中轴线的一系列庭院所组成，鲜有例外。这个原则同样适用于宗教和世俗建筑。从平面来说，庙宇与住宅并无基本的不同。因此，古代常有一些达官巨贾"舍宅为寺"。

佛教传入以前和石窟中所见的
木构架建筑之佐证

间接资料中的佐证

　　前章曾提到过的安阳附近的殷墟仅仅是一处已毁的遗址（图10）。当时的中国结构体系基本上与今日相同这个结论，是我们通过推论得出的。其证明还在于晚些时间的实例。反映木构架建筑面貌的最早资料，是战国时期（公元前403—前221年）一尊青铜钫上的雕饰（图9）。图为一座台基上的一栋二层楼房，有柱、出檐屋顶、门和栏杆。从结构的角度看，其平面布局应与殷墟遗址基本相同。特别重要的是图中表现出柱端具有斗栱这一特征，后来斗栱形成了严格的比例关系，很像欧洲建筑中的"柱式"。除这个雏形和殷墟遗址，以及其他几个刻画在铜器及漆器上的不那么重要也不说明问题的图像之外，公元前的中国建筑究竟是怎样的，人们还很不清楚。今后的考古发掘是否有可能把如此远古时期的中国建筑上部结构的面貌弄清楚，很令人怀疑。

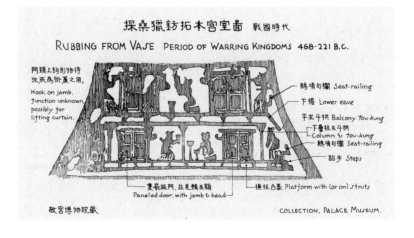

图 9
采桑猎钫拓本宫室
图（战国时代）

9
Early pictorial representation of a Chinese house

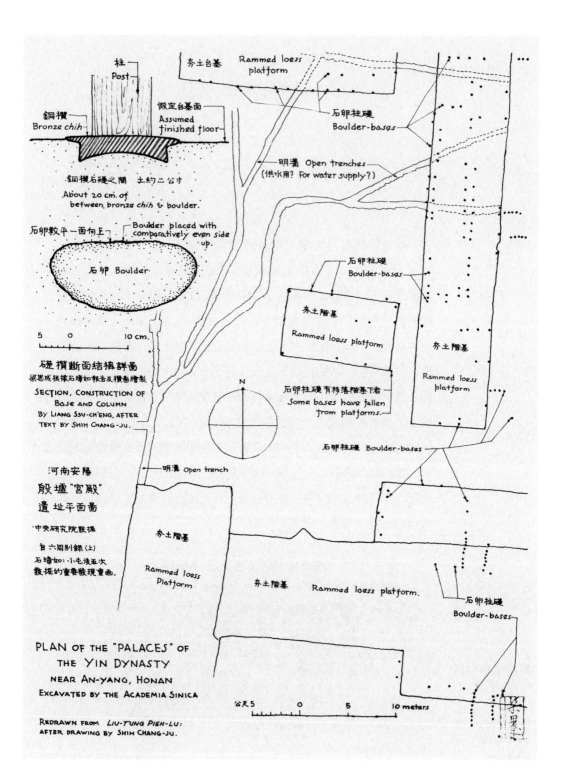

柱 Post

銅欖 Bronze chih

假定台基面 Assumed finished floor-

夯土台基 Rammed loess platform

石卵柱礎 Boulder-bases

銅欖石礎之間 土約二公寸
About 20 cm. of between bronze chih & boulder.

明潇 Open trenches (供水用? For water supply?)

石卵較平一面向上
Boulder placed with comparatively even side up.

石卵 Boulder

石卵柱礎 Boulder-bases

夯土階基 Rammed loess platform

5   0   10 cm.

礎欖斷面結構詳圖
梁思成根據石璋如報告及欖圖繪製
SECTION, CONSTRUCTION OF BASE AND COLUMN BY LIANG SSU-CH'ENG, AFTER TEXT BY SHIH CHANG-JU.

石卵柱礎有移著階基下者
Some bases have fallen from platforms.

N

石卵柱礎 Boulder-bases

明潇 Open trench

河南安陽
殷墟"宮殿"
遺址平面圖

中央研究院發掘

自六同別錄(上)
石璋如:小屯後五次
發掘的重要發現童畫.

夯土階基 Rammed loess Plattform

夯土階基 Rammed loess platform.

石卵柱礎 Boulder-bases

PLAN OF THE "PALACES" OF THE YIN DYNASTY
NEAR AN-YANG, HONAN
EXCAVATED BY THE ACADEMIA SINICA

公尺5   0   5   10 meters

REDRAWN FROM LIU-TUNG PIEH-LU:
AFTER DRAWING BY SHIH CHANG-JU.

26

汉代的佐证

在建筑上真正具有重要意义的最早遗例，见于东汉时期（公元25—220年）的墓。它们大体可分为三类：（1）崖墓（图11），其中一些具有高度的建筑性。这种墓大多见于四川省，少数见于湖南省；（2）独立的碑状纪念物——阙（图12），多数成双，位于通向宫殿、庙宇或陵墓的大道入口处的两侧；（3）供祭祀用的小型石屋——石室（图13），一般位于坟丘前。这些遗例都是石造，却如此真实地表现出斗栱和梁柱结构的基本特征，以致我们对其用为蓝本的木构建筑可以获得一个相当清楚的概念。

从上述三种遗例中，可以看出这一时期建筑的某些突出的特征：（1）柱呈八角形，冠以一个巨大的斗，斗下常有一条带状线道，代表皿板，即一方形小板；[1]（2）栱呈S字形，看来这不像是当时实有的木材成形方式；（3）屋顶和檐都由椽支承，上面覆以筒瓦，屋脊上有瓦饰。

汉代的木构宫殿和房屋的任何实物，现在都已不存。今天我们只能从当时的诗、文中略知其宏伟规模。但从墓中随葬的陶制明器建筑物（图14）以及陵墓石室壁上的画像石（图15）中，不难对汉代居住建筑有所了解。其中既有多层的大厦，也有简陋的普通民居。有一个实例的侧立面呈L字形，并以墙围成一个庭院（图16）。我们甚至从一座望楼中看出了佛塔的雏形。有些模型清楚地表现了建筑的木构架体系。这里我们再次看到了斗栱的主导作用。其作用是如此之大，以致如果没有对中国建筑中的这个决定性成分的透彻

[1]
6或7世纪以后，这一构件在中国建筑中即不再见。"皿板"一词借自日语，为7—8世纪日本建筑中该构件之名。——费慰梅注

图10
河南安阳殷墟"宫殿"遗址

10

Indications of Shang-Yin period architecture

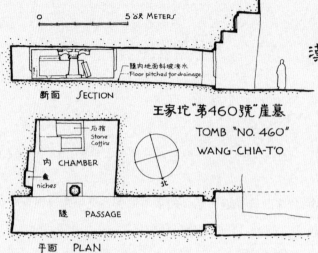

0   5 OR METERS

隧内地面斜坡瀉水
Floor pitched for drainage.

断面 SECTION

石棺
Stone
Coffins

内 CHAMBER

龕
niches

隧 PASSAGE

平面 PLAN

北

王家坨"第460號"崖墓

TOMB "NO. 460"
WANG-CHIA-T'O

四川彭山縣江口鎮附近

漢崖墓建築及彫飾

選自國立中央博物院
陳明達 "彭山崖墓報告"
未刊稿

ROCK-CUT TOMBS
NEAR CHIANG-K'OU,
P'ENG-SHAN HSIEN,
SZE-CH'UAN.
FROM CH'EN, M.-T.
"REPORT ON THE ROCK-CUT
TOMBS OF P'ENG-SHAN."
NATIONAL CENTRAL MUSEUM
(UNPUBLISHED)

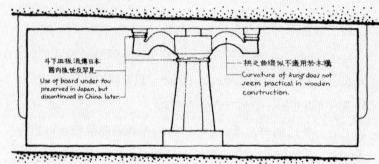

斗下皿板流傳日本
國內後世反罕見
Use of board under tou
preserved in Japan, but
discontinued in China later.

拱之曲綫似不適用於木構
Curvature of kung does not
seem practical in wooden
construction.

"第460號"
墓室及斗栱詳圖

DETAIL OF FUNERARY-
CHAMBER & "ORDER"
TOMB "Nº 460"

0   1 OR M.

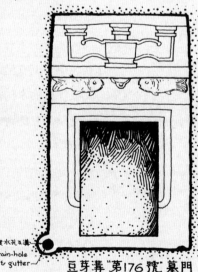

江口鎮"第355號"墓門

TOMB "Nº 355", CHIANG-K'OU, DETAIL OF ENTRANCE

瀉水孔及溝
Drain-hole
& gutter

豆芽溝"第176號"墓門

ENTRANCE, TOMB "Nº 176, TOU-YA-KOU.

了解，对中国建筑的研究便无法进行。从这些明器和画像石中，我们也可见到后世所用的所有五种屋顶结构：庑殿、硬山、悬山、歇山、攒尖（图3）。当时筒瓦的作用已和今天一样普遍了。

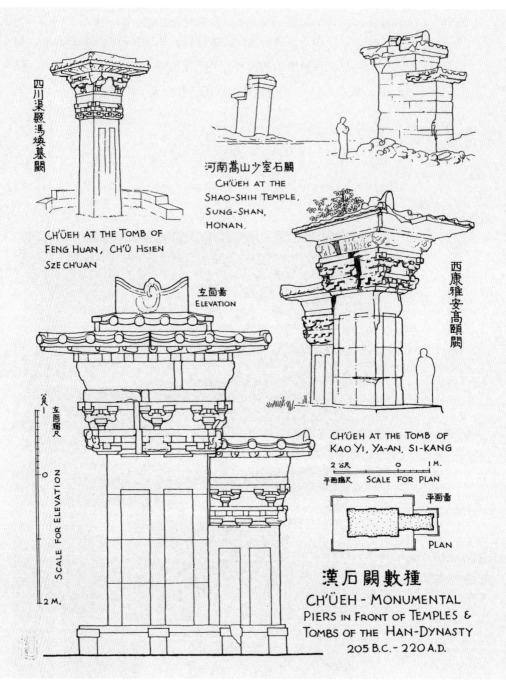

四川渠縣馮煥墓闕

CH'ÜEH AT THE TOMB OF
FENG HUAN, CH'Ü HSIEN
SZE CH'UAN

河南嵩山少室石闕
CH'ÜEH AT THE
SHAO-SHIH TEMPLE,
SUNG-SHAN,
HONAN.

左面圖
ELEVATION

西康雅安高頤闕

CH'ÜEH AT THE TOMB OF
KAO YI, YA-AN, SI-KANG

2 公尺   0   1 M.
平面縮尺   SCALE FOR PLAN

平面圖
PLAN

左面縮尺
SCALE FOR ELEVATION

漢石闕數種
CH'ÜEH - MONUMENTAL
PIERS IN FRONT OF TEMPLES &
TOMBS OF THE HAN-DYNASTY
205 B.C. - 220 A.D.

图 12
汉石阙　阙是对当
时简单的木构建筑
的模仿

12
Han stone *ch'üeh*.
These piers imitate
contemporary
simple wood
construction

图 13
独立式汉墓石室
（右页）

13
Han free-standing
tomb shrines

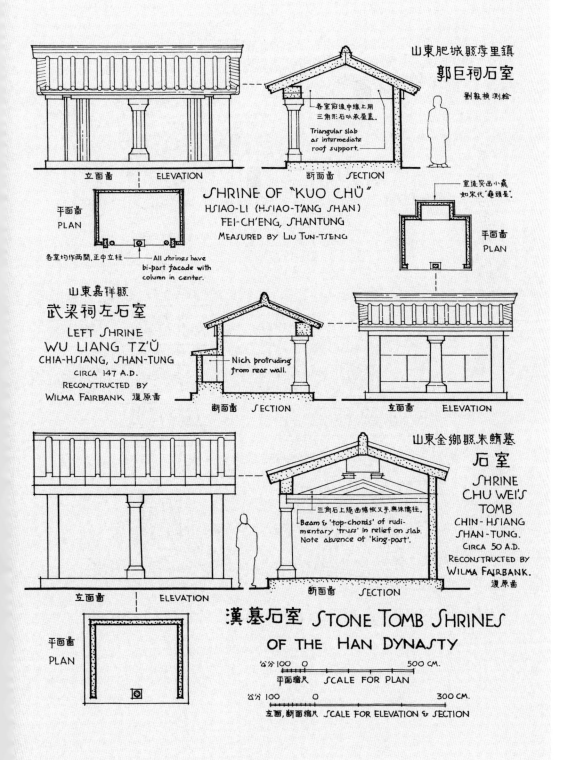

立面畫 ELEVATION   斷面畫 SECTION

山東肥城縣孝里鎮
郭巨祠石室

劉敦楨測繪

各室前邊中線上用
三角形石以承屋蓋.
Triangular slab
as intermediate
roof support.

平面畫 PLAN

各室均作兩間,正中立柱   All shrines have
bi-part facade with
column in center.

SHRINE OF "KUO CHÜ"
HSIAO-LI (HSIAO-T'ANG SHAN)
FEI-CH'ENG, SHANTUNG
MEASURED BY LIU TUN-TSENG

室後突出小龕
如宋代"龜頭屋".

平面畫 PLAN

山東嘉祥縣
武梁祠左石室

LEFT SHRINE
WU LIANG TZ'Ŭ
CHIA-HSIANG, SHAN-TUNG
CIRCA 147 A.D.
RECONSTRUCTED BY
WILMA FAIRBANK 復原畫

Nich protruding
from rear wall.

斷面畫 SECTION   立面畫 ELEVATION

立面畫 ELEVATION

山東金鄉縣朱鮪墓
石室

SHRINE
CHU WEI'S
TOMB
CHIN-HSIANG
SHAN-TUNG.
CIRCA 50 A.D.
RECONSTRUCTED BY
WILMA FAIRBANK.
復原畫

三角石上隱出檐枕叉手,無侏儒柱.
Beam & 'top-chords' of rudi-
mentary 'truss' in relief on slab.
Note absence of 'king-post'.

平面畫 PLAN

斷面畫 SECTION

漢墓石室 STONE TOMB SHRINES
OF THE HAN DYNASTY

公分 100 0   500 CM.
平面縮尺 SCALE FOR PLAN

公分 100 0   300 CM.
立面,斷面縮尺 SCALE FOR ELEVATION & SECTION

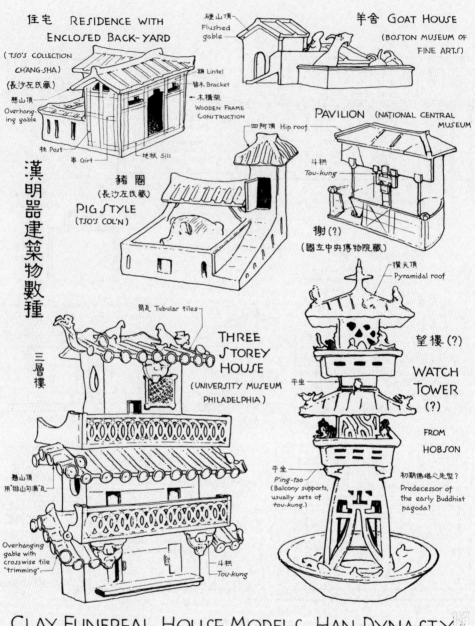

住宅 RESIDENCE WITH ENCLOSED BACK-YARD

(TSO'S COLLECTION CH'ANG-SHA)

(長沙左氏藏)

懸山頂 Overhang-ing gable

柱 Post

串 Girt

地栿 Sill

颗 Lintel

替木 Bracket

木構架 WOODEN FRAME CONSTRUCTION

硬山頂 Flushed gable

羊舍 GOAT HOUSE

(BOSTON MUSEUM OF FINE ARTS)

漢明器建築物數種

豬圈

(長沙左氏藏)

PIG STYLE

(TSO'S COL'N)

四阿頂 Hip roof

PAVILION (NATIONAL CENTRAL MUSEUM

斗拱 Tou-kung

榭(?)

(國立中央博物院藏)

三層樓

THREE STOREY HOUSE

(UNIVERSITY MUSEUM PHILADELPHIA)

筒瓦 Tubular tiles

懸山頂 用"排山勾滴瓦

Overhanging gable with crosswise tile "trimming"

斗拱 Tou-kung

攢尖頂 Pyramidal roof

望樓(?)

WATCH TOWER (?)

FROM HOBSON

平坐 Ping-tso

平坐 P'ing-tso (Balcony supports, usually sets of tou-kung.)

初期佛塔之先型? Predecessor of the early Buddhist pagoda?

CLAY FUNEREAL HOUSE MODELS, HAN DYNASTY

图 14　汉明器建筑物数种

14　Clay house models from Han tombs

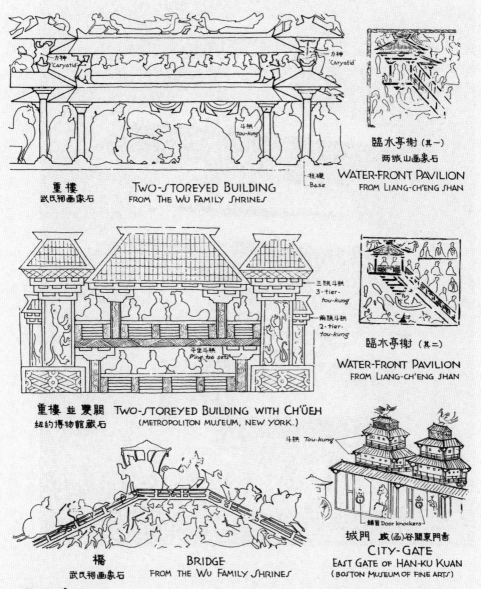

力神 'Caryatid'
力神 'Caryatid'
斗栱 Tou-kung
柱礎 Base

重樓
武氏祠画像石
TWO-STOREYED BUILDING
FROM THE WU FAMILY SHRINES

臨水亭榭 (其一)
兩城山画象石
WATER-FRONT PAVILION
FROM LIANG-CH'ENG SHAN

三跳斗栱
3-tier-tou-kung
兩跳斗栱
2-tier-tou-kung
平坐斗栱
Ping-tso sets

臨水亭榭 (其二)
WATER-FRONT PAVILION
FROM LIANG-CH'ENG SHAN

重樓 並 雙闕
紐約博物館藏石
TWO-STOREYED BUILDING WITH CH'ÜEH
(METROPOLITON MUSEUM, NEW YORK.)

斗栱 Tou-kung

橋
武氏祠画象石
BRIDGE
FROM THE WU FAMILY SHRINES

鋪首 Door knockers

城門 威 (函) 谷關東門高
CITY-GATE
EAST GATE OF HAN-KU KUAN
(BOSTON MUSEUM OF FINE ARTS)

漢画象石中
建築數種
ARCHITECTURE FOUND IN ENGRAVED STONES
(OR RELIEFS) OF THE HAN DYNASTY 205 B.C.-220 A.D.

图 15　　　　　15
汉画像石中建筑　　Architecture
数种　　　　　　　depicted in Han
　　　　　　　　　engraved reliefs

图 16

汉代陶制明器　美国密苏里州堪萨斯市纳
尔逊 – 阿特金斯博物馆藏。这座三层的汉
代住宅的结构以粗略的造型和彩绘来表现。
不仅在檐下，而且在阳台下面都用了斗栱。
在第三层使用了转角斗栱，但没有解决问
题。大门两侧的角楼与华北地区所见石阙
相似

16

Han funerary clay house model, Nelson-
Atkins Museum, Kansas City, Missouri.
The structural frame of a three-story Han
dwelling is indicated by crude modeling and
painting. *Tou-kung* are used not only under
the eaves but also under the balcony. Corner
bracket sets are introduced on the third story,
but the solution is not satisfactory. The two
corner towers flanking the gate are similar to
the stone *ch'üeh* found in North China

石窟中的佐证

佛教传入中国的时期，大体上相当于公元开始的时候。虽然根据记载，早在 3 世纪初中国就已出现了"下为重楼，上累金盘"的佛塔，但现存的佛教建筑却都是 5 世纪中叶以后的实物。从此时起直到 14 世纪晚期，中国建筑的历史几乎全是佛教（以及少数道教）庙宇和塔的历史。

山西大同近郊的云冈石窟（5 世纪中叶至 5 世纪末），虽无疑渊源于印度，其原型来自印度卡尔里［Karli］、阿旃陀［Ajanta］等地，但其发源地对它的影响却小得惊人。石窟的建筑手法几乎完全是中国式的。唯一标志着其外来影响的，就是建造石窟这种想法本身，以及其希腊—佛教型的装饰花纹，如莨苕、卵箭、卍字、花绳和莲珠等等。从那时起，它们在中国装饰纹样的语汇中生了根，并大大地丰富了中国的饰纹（图 17）。

我们可以从两方面来研究云冈石窟的建筑：（1）研究石窟本身，包括其内外建筑手法；（2）从窟壁浮雕所表现的建筑物上研究当时的木构和砖石建筑（图 18）。浮雕中有许多殿堂和塔的刻像，这些建筑当时曾遍布于华北和华中的平原和山区。

在崖石上开凿石窟的做法在华北地区一直延续到唐（公元618—907 年）中叶，此后，在西南，特别是四川省，直到明代（公元 1368—1644 年）还有这种做法。其中只有早期的石窟才引起史家的兴趣。山西太原附近的天龙山石窟和河南、河北两省交界处的响堂山石窟最富建筑色彩，它们都是北齐和隋代（6 世纪末至 7

世纪初）的遗迹（图19）。

这些石窟以石刻保存了当时的木构建筑的逼真摹本。其最显著特色是：柱大多呈八角形，柱头作大斗状，同汉崖墓中所见相似。柱头上置阑额，阑额再承铺作中的栌斗。后来，这种做法演变为将阑额直接卯合于柱端，而把铺作中的栌斗直接放在柱顶上（两个斗合并为一）。

在石窟所表现的建筑手法中，斗栱始终是一个主导的构件。它们仍如汉崖墓中所见的那么简单，但S形的已经取直，似更合理。在两柱之间的阑额上，使用了人字形补间铺作。这种做法在现存的中国建筑中仅存一例，即河南登封县会善寺净藏禅师塔（图64d，图64e），其上尚有用砖模仿的人字形补间铺作。塔建于公元746年［唐天宝五载］。此外只能在建于这一时期的几座日本木构建筑上见到。

图 17
云冈石窟中一座门
的饰纹细部
17
Detail of an interior
doorway. Yun-kang
Caves, near Ta-t'ung,
Shansi, 450–500

图 18
云冈石窟所表现之
北魏建筑（右页）
18
Architectural
elements carved in
the Yun-kang Caves

图像中国建筑史

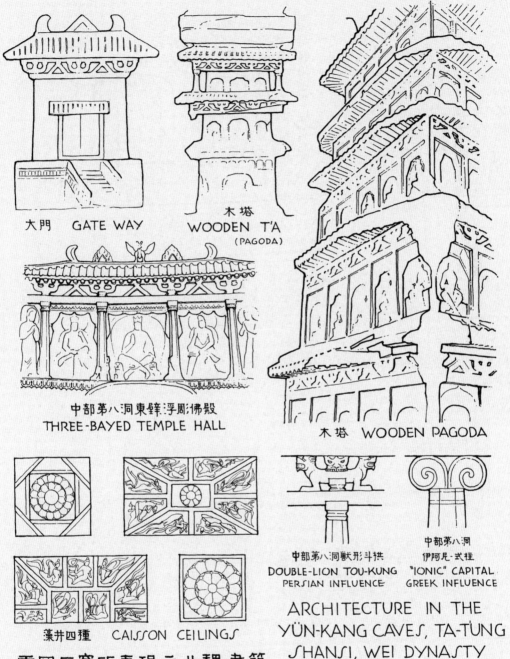

大門　GATE WAY

木塔
WOODEN T'A
(PAGODA)

中部第八洞東鎊浮彫佛殿
THREE-BAYED TEMPLE HALL

木塔　WOODEN PAGODA

藻井四種　CAISSON CEILINGS

中部第八洞獸形斗拱
DOUBLE-LION TOU-KUNG
PERSIAN INFLUENCE

中部第八洞
伊阿尼式柱
"IONIC" CAPITAL
GREEK INFLUENCE

雲岡石窟所表現之北魏建築

ARCHITECTURE IN THE
YÜN-KANG CAVES, TA-TUNG
SHANSI, WEI DYNASTY
EXECUTED BETWEEN 450 & 500 A.D.

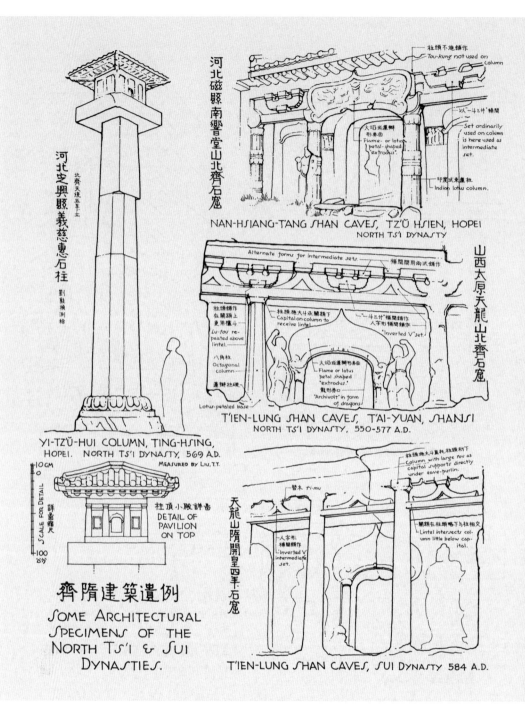

河北磁縣南響堂山北齊石窟

河北芝興縣義慈惠石柱

北齊天統五年立

劉敦楨測繪

- 柱頭不施鋪作
  Tou-kung not used on Column

- 以一斗二升'補間
  Set ordinarily used on column is here used as intermediate set.

- 大焰或蓮瓣形券面
  Flame- or lotus- petal-shaped "extrados."

- 印度或束蓮柱
  Indian lotus column.

NAN-HSIANG-TANG SHAN CAVES, TZ'Ŭ HSIEN, HOPEI
NORTH TS'I DYNASTY

山西大原天龍山北齊石窟

Alternate forms for intermediate sets.

- 補間關用兩式鋪作

- 柱頭應大斗承闌額下
  Capital on column to receive lintel

- 一斗三升'補間鋪作 人字形補間鋪作
  "Inverted V" set.

- 柱頭鋪作 在闌額上 更用櫨斗
  Lu-tou re- peated above lintel.

- 八角柱
  Octagonal column

- 大焰或蓮瓣形券面
  Flame or lotus petal shaped "extrados."

- 龍形券身
  "Archivolt" in form of dragons

- 蓮瓣柱礎

- Lotus-petaled base

T'IEN-LUNG SHAN CAVES, T'AI-YUAN, SHANSI
NORTH TS'I DYNASTY, 550-577 A.D.

YI-TZ'Ŭ-HUI COLUMN, TING-HSING,
HOPEI. NORTH TS'I DYNASTY, 569 A.D.
MEASURED BY LIU, T.T.

10CM 0
SCALE FOR DETAIL
詳高應尺
100
8分

DETAIL OF PAVILION ON TOP
柱頂小殿詳高

天龍山隋開皇四年石窟

- 替木 ti-mu

- 人字柱 補間鋪作 "Inverted V intermediate set.

- 柱頭應大斗直柱柱頭枋下
  Column, with large tou as capital supports directly under eave-purlin.

- 闌額在柱頭略下與柱相交
  Lintel intersects col- umn little below cap- ital.

齊隋建築遺例
SOME ARCHITECTURAL
SPECIMENS OF THE
NORTH TS'I & SUI
DYNASTIES.

T'IEN-LUNG SHAN CAVES, SUI DYNASTY 584 A.D.

图 19   19
齐隋建筑遗例   Architectural
                       representations
                       from the North Ch'i
                       and Sui dynasties

木构建筑重要遗例

中国人所用的主要建筑材料——木材，是非常容易朽坏的。它们会遭到风雨和蛀虫的自然侵蚀，又极易燃烧。在宗教建筑中，它们又总是受到善男信女所供奉的香火的威胁。加之，时时的内战和宗教斗争也很不利于木构建筑的保存。每一新朝代的开国者，依惯例总是要对败者的都城大肆劫掠，他们不是造反者，就是军阀或北方落后民族的首领。怀着对被征服的原统治者极大的敌意，他们总是要把大大小小的王公贵戚们那无数金碧辉煌的宫殿夷为一片废墟（作为这种野蛮习惯的极少数例外之一的，是 1912 年中华民国的建立。当时，清朝皇宫作为一处博物院，向公众开放了）。

　　尽管中国一直被认为是个宗教自由的国家，但自 5 世纪至 9 世纪，至少曾发生过三次对佛教的大迫害。其中第三次发生在公元 845 年［唐武宗会昌五年，"会昌灭法"］，当时全国的佛教庙宇寺院几乎被扫荡一空。可能正是这些情况以及木材的易毁性，说明了何以中国 9 世纪中叶以前的木构建筑已完全无存。

　　中国近年来的趋势，特别是自中华民国建立以来，对于古建筑的保存来说仍是不利的。自 19 世纪中叶以来，中国屡败于近代列强，使中国的知识分子和统治阶级对于一切国粹都失去了信心。他们的审美标准全被搅乱了：古老的被抛弃了；对于新的，即西方的，却又茫然无所知。佛教和道教被斥为纯粹的迷信，而且，不无理由地被视为使中国人停滞的原因之一。总的倾向是反对传统观念。许多庙宇被没收并改作俗用，被反对传统的官员们用作学校、办公室、谷仓，甚至成了兵营、军火库和收容所。在最好的情况下，这些房子被改建以适应其新功用；而最坏的，这些倒霉的建筑物竟成了毫无纪律、薪饷不足的大兵们任意糟蹋的牺牲品，他们由于缺少燃料，常把一切可拆的部件——槅扇、门、窗、栏杆，甚至斗栱都拆下来烧火做饭。

直到 20 年代后期，中国的知识分子才开始认识到中国自己的建筑艺术的重要性绝不低于其书法和绘画。首先，有一些外国人建造了一批中国式的建筑；其次，西方和日本的学者出版了一些书和文章来论述中国建筑；最后，有一批到西方学习建筑技术的中国留学生回到了国内，他们认识到建筑不仅是一些砖头和木料而已，它是一门艺术，是民族和时代的表征，是一种文化遗产。于是，知识阶层过去对于"匠作之事"的轻视态度，逐渐转变为赞赏和钦佩。但是要想使地方当局也获得这种认识，可不是容易的事，而保护古迹却有赖于这些人。在那些无知者和漠不关心者手中，中国的古建筑仍在不断地遭到破坏。

最后，日本对中国的侵略战争（1937—1945 年）究竟造成了多大的破坏，目前尚不可知。如果本书所提到的许多文物建筑今后将仅仅留下这些照片和图版而原物不复存在，那也是预料中事。

对于现存的，更确切地说 30 年代尚存的这些建筑，我们可试分为三个主要时期："豪劲时期""醇和时期"和"羁直时期"（图20，图 21）。

豪劲时期包括自 9 世纪中叶至 11 世纪中叶这一时期，即自唐宣宗大中至宋仁宗天圣末年。其特征是比例和结构的壮硕坚实，这是繁荣的唐代必然的特色，而我们所提到的这一时期仅是一个光辉的尾声而已。

醇和时期自 11 世纪中叶至 14 世纪末，即自宋英宗治平，中经元代，至明太祖洪武末。其特点是比例优雅、细节精美。

羁直时期系自 15 世纪初到 19 世纪末，即自明成祖（永乐）年间夺取其侄帝位，由南京迁都北京，一直延续到清王朝被中华民国

图 20
历代木构殿堂外观演变图
20
Evolution of the general appearance of timber-frame halls

歷代木構殿堂外觀演變圖

EVOLUTION OF THE GENERAL APPEARANCE OF TIMBER-FRAMED HALLS

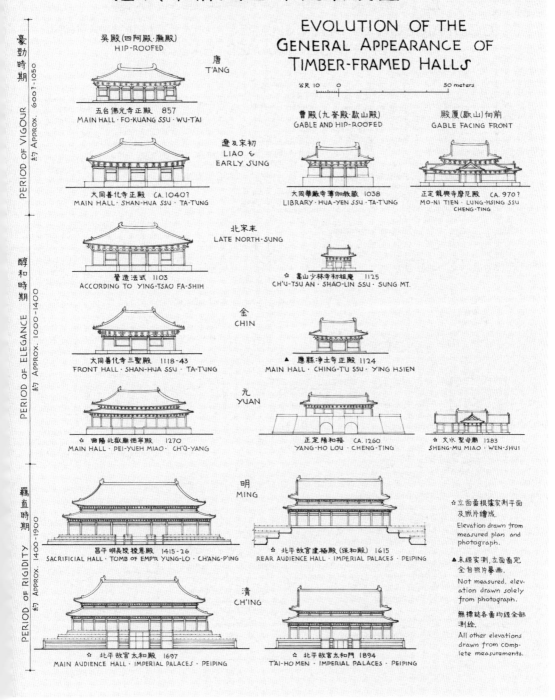

公尺 10 · 0 ——————— 50 meters

豪勁時期 · PERIOD OF VIGOUR · 約 Approx. 600?-1050

吳殿(四阿殿·廡殿)
HIP-ROOFED

唐
T'ANG

五台佛光寺正殿 857
MAIN HALL · FO-KUANG SSU · WU-T'AI

遼及宋初
LIAO & EARLY SUNG

大同善化寺正殿 CA.1040?
MAIN HALL · SHAN-HUA SSU · TA-T'UNG

曹殿(九脊殿·歇山殿)
GABLE AND HIP-ROOFED

大同華嚴寺薄伽教藏 1038
LIBRARY · HUA-YEN SSU · TA-T'UNG

殿廈(歇山)向前
GABLE FACING FRONT

正定龍興寺摩尼殿 CA.970?
MO-NI TIEN · LUNG-HSING SSU CHENG-TING

醇和時期 · PERIOD OF ELEGANCE · 約 Approx. 1000-1400

北宋末
LATE NORTH·SUNG

營造法式 1103
ACCORDING TO YING-TSAO FA-SHIH

✿ 嵩山少林寺初祖庵 1125
CH'U-TSU AN · SHAO-LIN SSU · SUNG MT.

金
CHIN

大同善化寺三聖殿 1118-43
FRONT HALL · SHAN-HUA SSU · TA-T'UNG

▲ 應縣淨土寺正殿 1124
MAIN HALL · CHING-T'U SSU · YING HSIEN

元
YUAN

✿ 曲陽北嶽廟德寧殿 1270
MAIN HALL · PEI-YUEH MIAO · CH'Ü-YANG

正定陽和樓 CA.1260
YANG-HO LOU · CHENG-TING

✿ 文水聖母廟 1283
SHENG-MU MIAO · WEN-SHUI

羈直時期 · PERIOD OF RIGIDITY · 約 Approx. 1400-1900

明
MING

✿ 昌平明長陵棱恩殿 1415-26
SACRIFICIAL HALL · TOMB OF EMP'R YUNG-LO · CH'ANG-P'ING

✿ 北平故宮建極殿(保和殿) 1615
REAR AUDIENCE HALL · IMPERIAL PALACES · PEIPING

清
CH'ING

✿ 北平故宮太和殿 1697
MAIN AUDIENCE HALL · IMPERIAL PALACES · PEIPING

✿ 北平故宮太和門 1894
T'AI-HO MEN · IMPERIAL PALACES · PEIPING

✿ 立面畫根據實測平面及照片繪成。
Elevation drawn from measured plan and photograph.

▲ 未經實測,立面畫完全自照片臆畫。
Not measured, elevation drawn solely from photograph.

無標誌各畫均經全部測繪。
All other elevations drawn from complete measurements.

43

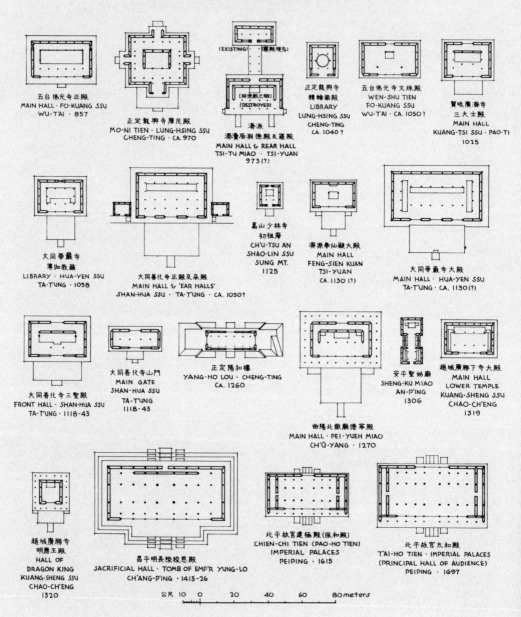

歷代殿堂平面及列柱位置比較圖
COMPARISON OF PLAN SHAPES AND COLUMNIATION OF TIMBER-FRAMED HALLS

五台佛光寺正殿
MAIN HALL · FO-KUANG SSU
WU-T'AI · 857

正定龍興寺摩尼殿
MO-NI TIEN · LUNG-HSING SSU
CHENG-TING · CA. 970

(EXISTING) (慶殿現存)
(順德殿己毀)
(DESTROYED)
濟源
濟源廟淵德殿及寢殿
MAIN HALL & REAR HALL
TSI-TU MIAO · TSI-YUAN
973 (?)

正定龍興寺
轉輪藏殿
LIBRARY
LUNG-HSING SSU
CHENG-TING
CA. 1040 ?

五台佛光寺文殊殿
WEN-SHU TIEN
FO-KUANG SSU
WU-T'AI · CA. 1050?

寶坻廣濟寺
三大士殿
MAIN HALL
KUANG-TSI SSU · PAO-TI
1025

大同華嚴寺
薄伽教藏
LIBRARY · HUA-YEN SSU
TA-T'UNG · 1038

大同善化寺正殿及朵殿
MAIN HALL & 'EAR HALLS'
SHAN-HUA SSU · TA-T'UNG · CA. 1050?

嵩山少林寺
初祖庵
CH'U-TSU AN
SHAO-LIN SSU
SUNG MT.
1125

濟源奉仙觀大殿
MAIN HALL
FENG-SIEN KUAN
TSI-YUAN
CA. 1130 (?)

大同華嚴寺大殿
MAIN HALL · HUA-YEN SSU
TA-T'UNG · CA. 1130(?)

大同善化寺三聖殿
FRONT HALL · SHAN-HUA SSU
TA-T'UNG · 1118-43

大同善化寺山門
MAIN GATE
SHAN-HUA SSU
TA-T'UNG
1118-43

正定陽和樓
YANG-HO LOU · CHENG-TING
CA. 1260

曲陽北嶽廟德寧殿
MAIN HALL · PEI-YUEH MIAO
CH'Ü-YANG · 1270

安平聖姑廟
SHENG-KU MIAO
AN-P'ING
1306

趙城廣勝下寺大殿
MAIN HALL
LOWER TEMPLE
KUANG-SHENG SSU
CHAO-CH'ENG
1319

趙城廣勝寺
明應王殿
HALL OF
DRAGON KING
KUANG-SHENG SSU
CHAO-CH'ENG
1320

昌平明長陵祾恩殿
SACRIFICIAL HALL · TOMB OF EMP'R YUNG-LO
CH'ANG-P'ING · 1415-26

北平故宮建極殿(保和殿)
CHIEN-CHI TIEN (PAO-HO TIEN)
IMPERIAL PALACES
PEIPING · 1615

北平故宮太和殿
T'AI-HO TIEN · IMPERIAL PALACES
(PRINCIPAL HALL OF AUDIENCE)
PEIPING · 1697

公尺 10 0   20   40   60   80 meters

图 21
历代殿堂平面及列
柱位置比较图
21
Comparison of plan
and columniation
of timber-frame
halls

推翻；这一时期的特点是建筑普遍趋向僵硬；由于所有水平构件尺
寸过大而使建筑比例变得笨拙；以及斗栱（相对于整个建筑来说）
尺寸缩小，因而补间铺作攒数增加，结果竟失去其原来的结构功能
而蜕化为纯粹的装饰品了。

这样的分期法当然只是我个人的见解。在一种演化的过程中，
不可能将那些难以觉察的进程截然分开。因此，在一座早期的建筑
中，也可能见到某些后来风格或后来特点的前兆；而在远离文化政
治中心的边远地区的某个晚期建筑上，也会发现仍有一些早已过时
的传统依然故我。不同时期的特征必然会有较长时间的互相交错。

豪劲时期

（约公元 850—1050 年）

间接资料中的佐证

　　豪劲时期的建筑，今天只有属于其末期的少数实物尚存。这个
时期在 9 世纪中叶前必定有其相当长的一段光辉历史，甚至远溯到
唐朝初年，即 7 世纪早期。然而对于这个时期中假定的前半段的木
构建筑，我们却只能借助于当时的图像艺术去探索其消息。在陕西
西安大雁塔（公元 701—704 年）［唐武后长安间］西门门楣上的一
幅石刻上，有一个佛寺大殿的细致而准确的图形（图 22）。从中人们
第一次看到斗栱中有向外挑出的华栱，作悬臂用以支承上面那个很
深的出檐。但这并不是说，华栱直到此时才出现。相反，它必定早
已被使用，甚至可能已经几百年了，所以才能发展到如此适当而成
熟的程度，并被摹写于画图之中，成为一种可被我们视为典型的形
象。两层外挑的华栱上置以横栱，还使用了人字栱作为补间铺作。
正脊两端的鸱尾和垂脊上面花蕾形的装饰都与后世的不同。图中唯
一不够准确之处是柱子过细，可能是因为不愿挡住殿内的佛像和罗
汉才这么画的。

　　另一些重要的资料是由斯坦因爵士和伯希和教授取自甘肃敦煌
千佛洞（5—11 世纪）后加以复原的唐代绘画，它们现存英国不列
颠博物馆和法国卢浮宫。这些绢画和壁画所描绘的是西方极乐世
界，其中有许多建筑物如殿、阁、亭、塔之类的形象。画中的斗栱
不仅有外挑的华栱，还有斜置的、带尖端的昂。这是利用杠杆原理
来支承深远出檐。这种结构方式，后来在大多数殿堂建筑中都可见

A TEMPLE HALL OF THE T'ANG DYNASTY

AFTER A RUBBING OF THE ENGRAVING ON THE TYMPANIUM OVER THE WEST
GATEWAY OF TA-YEN T'A, TZ'U-EN SSŬ, SI-AN, SHENSI

唐代佛殿圖　摹自陕西长安大雁塔西門門楣石画像

图 22

唐代佛殿图（摹自
陕西西安大雁塔西
门门楣石画像）

22

A temple hall of
the T'ang dynasty,
engraved relief
from the Ta-yen
T'a（Wild Goose
Pagoda）, Sian
Shensi, 701–704

〔1〕
后来在同一地区发
现了一座更早的
较小而较简单的
殿——建于 782 年的
南禅寺。见 1954 年
第一期《文物》杂
志。——费慰梅注

到。其他许多建筑细节在这些绘画中都可以找到（图 23）。

在四川的某些晚唐石窟中，也可见到同样主题的浮雕，但其中的建筑比绘画中的要简单得多，显然是受材料所限。把表现同样主题的浮雕和绘画加以比较，使我们有理由推断，汉墓和魏、齐、隋代的石窟中所见的建筑，肯定只是对于原状已发展得远为充分的木构建筑的简化了的描绘而已。

佛光寺

目前所知的木构建筑中最早的实物，是山西五台山佛光寺的大殿[1]（图 24）。该殿建于公元 857 年 [唐宣宗大中十一年]，即会昌灭法之后十二年。该址原有一座七间、三层、九十五尺 [中国尺，此说系作者引自中国古文献] 高，供有弥勒巨像的大阁。现存大殿是被毁后重建的，为单层、七间，其严谨而壮硕的比例使人印象极深。巨大的斗栱共有四层伸出的臂 ["出跳"] ——两层华栱，两层昂 ["双杪

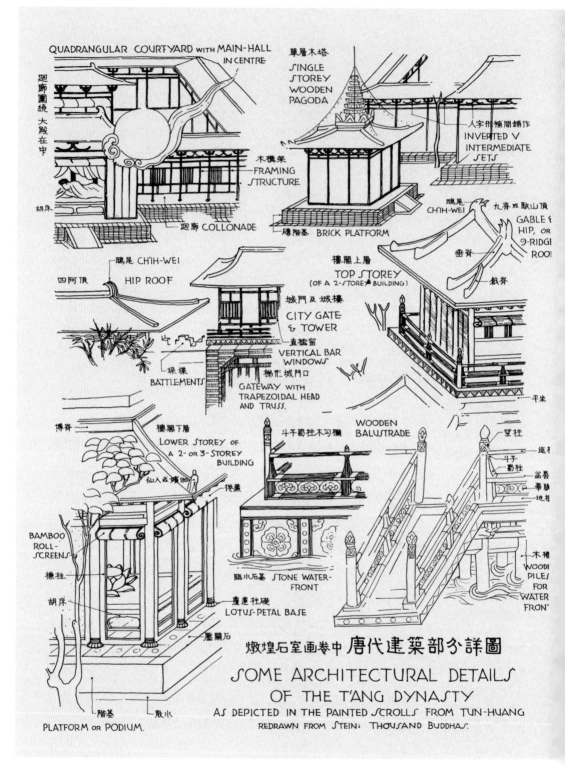

QUADRANGULAR COURTYARD with MAIN-HALL IN CENTRE

回廊圍繞 大殿在中

胡床

木構架 FRAMING STRUCTURE

迴廊 COLLONADE

單層木塔 SINGLE STOREY WOODEN PAGODA

人字形補間鋪作 INVERTED V INTERMEDIATE SETS

磚階基 BRICK PLATFORM

鴟尾 CH'IH-WEI

九脊或歇山頂 GABLE & HIP, OR 9-RIDGE ROOF

垂脊

戧脊

四阿頂 鴟尾 CH'IH-WEI HIP ROOF

樓閣上層 TOP STOREY (OF A 2-STOREY BUILDING)

城門及城樓 CITY GATE & TOWER

直櫺窗 VERTICAL BAR WINDOWS

堞堞 BATTLEMENTS

梯形城門口 GATEWAY with TRAPEZOIDAL HEAD AND TRUSS.

平坐

博脊 樓閣下層 LOWER STOREY of A 2- or 3-STOREY BUILDING

仙人或嬪伽

捲薰

斗子蜀柱不勾欄

WOODEN BALUSTRADE

望柱

巡杖

斗子 蜀柱

盆唇 華版 地栿

BAMBOO ROLL-SCREENS

欉柱

胡床

臨水石基 STONE WATER-FRONT

青蓮柱礎 LOTUS-PETAL BASE

木樁 WOODEN PILES FOR WATER FRONT

階基

散水

壓闌石

PLATFORM or PODIUM.

燉煌石室画卷中 唐代建築部分詳圖

SOME ARCHITECTURAL DETAILS OF THE T'ANG DYNASTY

AS DEPICTED IN THE PAINTED SCROLLS FROM TUN-HUANG

REDRAWN FROM STEIN: THOUSAND BUDDHAS.

双下昂"］，斗栱高度约等于柱高的一半，其中每一构件都有其结构功能，从而使整幢建筑显得非常庄重，这是后来建筑所未见的。

大殿内部显得十分典雅端庄。月梁横跨内柱间，两端各由四跳华栱支承，将其荷载传递到内柱上。殿内所有梁［明栿］的各面都呈曲线，与大殿庄严的外观恰成对照。月梁的两侧微凸，上下则略呈弓形，使人产生一种强劲有力的观感，而这是直梁所不具备的。

从结构演变阶段的角度看，这座大殿的最重要之处就在于有着直接支承屋脊的人字形构架；在最高一层梁的上面，有互相抵靠着的一对人字形叉手以撑托脊槫，而完全不用侏儒柱。这是早期构架方法留存下来的一个仅见的实例。过去只在山东金乡县朱鲔墓石室（公元 1 世纪）雕刻（图 13）和敦煌的一幅壁画中见到过类似的结构。其他实例，还可见日本奈良法隆寺庭院周围的柱廊。佛光寺是国内现存此类结构的唯一遗例。

尤为珍贵的是，这座大殿内还保存了一批与建筑物同时的塑像、壁画和题字。在巨大的须弥座上，有三十多尊巨型佛像和菩萨像。但最引人注目的，却是两尊谦卑的等身人像，其中之一为本殿女施主宁公遇像，另一为本寺住持愿诚和尚像，他是弥勒大阁在公元 845 年［会昌五年］被毁后主持重修的人。梁的下面有墨笔书写的大殿重修时本地区文武官员及施主姓名。在一处栱眼壁上留有一幅大小适中的壁画，为唐风无疑。与之相比，旁边内额上绘于公元 1122 年［宣和四年］的宋代壁画，虽然也十分珍贵，却不免逊色了。这样，在一座殿内竟保存了中国所有的四种造型艺术，而且都是唐代的，其中任何一件都足以被视为国宝。四美荟于一殿，真是不可思议的奇迹。

此后的一百二十年是一段空白，其间竟无一处木构建筑遗存下来。在这以后敦煌石窟中有两座年代可考的木建筑，分别建于公元 976 年［宋太平兴国元年］和公元 980 年［太平兴国五年］，但它们几乎难以被称为真正的建筑，而仅仅是石窟入口处的窟廊，然而毕竟是罕见的宋初遗物。

图 23
敦煌石室画卷中唐代建筑部分详图
23
T'ang architectural details from scrolls discovered in the Caves of the Thousand Buddhas, Tunhuang Kansu

图 24

山西五台山佛光寺
大殿

24

Main Hall, Fo-kuang
Ssu, Yu-t'ai Shan,
Shansi, 857

a. 全景。右上方为
大殿，左侧长屋顶
为后来建的文殊殿

a. General view.
The Main Hall is
at upper right; the
long roof at left is
the later Wen-shu
Tien

b. 立面

b. Facade

c. 外檐斗栱

c. Exterior *tou-kung*

d. 佛光寺大殿前
廊，前景中三脚架
旁立者为梁思成

d. Front interior
gallery. Liang
at tripod in for-
eground

e. 大殿内槽斗栱及
梁（左页上）
e. Inside of hy-
postyle

f. 佛光寺大殿内槽
斗栱及唐代壁画
（左页下）
f. Interior of *tou-kung*
and T'ang mural

g. 屋顶构架
g. Roof frame
h. 女施主宁公遇像
h. Statue of lady
donor
i. 愿诚和尚像
i. Statue of abbot

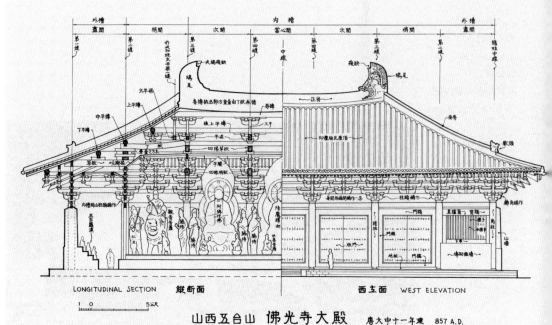

山西五台山 佛光寺大殿　唐大中十一年建　857 A.D.

MAIN HALL OF FO-KUANG SSU · WU-T'AI SHAN · SHANSI

LONGITUDINAL SECTION　縱斷面

WEST ELEVATION　西立面

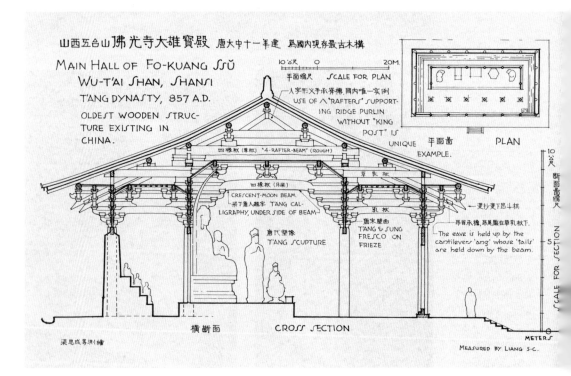

山西五台山 佛光寺大雄寶殿　唐大中十一年建　爲國內現存最古木構

MAIN HALL OF FO-KUANG SSŬ
WU-T'AI SHAN, SHANSI
T'ANG DYNASTY, 857 A.D.
OLDEST WOODEN STRUC-
TURE EXISTING IN
CHINA.

SCALE FOR PLAN　平面橫尺

人字形 叉手承脊榑，國內唯一實例
USE OF ∧ "RAFTERS" SUPPORT-
ING RIDGE PURLIN
WITHOUT "KING
POST" IS
UNIQUE
EXAMPLE.

PLAN　平面圖

"4-RAFTER-BEAM" (ROUGH)　四椽栿 (草栿)

四椽栿 (月樑)
CRESCENT-MOON BEAM
梁下入題字 T'ANG CAL-
LIGRAPHY, UNDER SIDE OF BEAM

草乳栿

乳栿

T'ANG SCUPTURE　唐代塑像

T'ANG & SUNG
FRESCO ON
FRIEZE　唐宋壁畫

昂抄昂下昂斗栱

昂抄承榑，昂尾壓在草乳栿下。
The eave is held up by the
cantilevers 'ang' whose 'tails'
are held down by the beam.

CROSS SECTION　橫斷面

SCALE FOR SECTION　斷面高縮尺

METERS

MEASURED BY LIANG S-C.

梁思成摹測繪

54

独乐寺的两栋建筑物

按年代顺序，再往后的木构建筑是河北蓟县独乐寺中宏伟的观音阁及其山门，同建于公元984年［辽统和二年］。当时这一地区正处于辽代契丹人的统治之下。阁（图25）为两层，中间夹有一个暗层。阁中有一尊十一面观音巨型塑像，高约五十二英尺［约十六米］，是国内同类塑像中最大的。阁的上面两层环像而建，中间形成一个空井，成为围绕像的胸部和臀部的两圈回廊。从结构上说，阁由三层"叠柱式"结构（斗栱、梁柱的构架相叠）组成，每层都有齐全的柱和斗栱。斗栱的比例和细部与佛光寺唐代大殿十分接近。但在此处，除在顶层采用了双层栱和双层昂结构（"双杪双下昂"）外，在平坐和下层外檐柱上还使用了没有昂的重栱。略似大雁塔门楣石刻中的形象（图22）。阁内斗栱，位置不同，形式各异，各司其职以支承整栋建筑，从而形成了一个斗栱的大展览。棋盘状小方格的天花［平闇］用直梁而不用月梁承托。脊槫的支承，用叉手外尚加侏儒柱，形成一个简单的桁架。在这以后一个时期中，侏儒柱逐渐完全取代了叉手，成了将脊槫的重量传递到梁上去的唯一构件。这样，叉手之有无，以及它们与侏儒柱相比的大小，就成了辨别建筑年代的一个明显标志。这时期的另一特征除罕例以外，是其内柱常与檐柱同高。梁架的上部由相叠的斗栱支承，而极少如后来那样，把内柱加高以接近高处的构件。

独乐寺的山门（图26），是一座不大的建筑物，檐下有简单的斗栱。从平面上看，这是一座典型的中国式大门。在它的长轴上有一排柱，两扇门即安装在柱上。内部结构是所谓"彻上露明造"，也就是没有天花，承托屋顶的结构构件都露在外面。山门展示了木作艺术的一个精巧实例；整个结构都是功能性的，但在外表上却极富装饰性。这种双重品质是中国建筑结构体系的最大优点所在。

在这两座建筑之后的三百年里，即辽、宋、金三代，有三十余座木构建筑遗存至今。尽管数量不多，年代分布也不均衡，但仍可将它们顺序排成一个没有间隙的系列。其中，约有二十余座属于我

j. 佛光寺大殿纵断面和西立面图
j. Elevation and longitudinal section
k. 佛光寺大殿平面和断面图
k. Plan and cross section

图 25

河北蓟县独乐寺观
音阁

25

Kuan-yin Ke，Tu-
le Ssu，Chi Hsien，
Hopei，984

a. 全景、立面，檐
角下立柱为后来
新加

a. General view,
facade（eave props
added later）

b. 模型，显示了细
部结构

b. Model, showing
structural details

c. 外檐斗栱

c. Exterior *tou-kung*

d. 内部斗栱

d. Interior well

e. 观音像仰视

e. Kuan-yin statue from below

f. 第三层内景

f. Interior, upper story

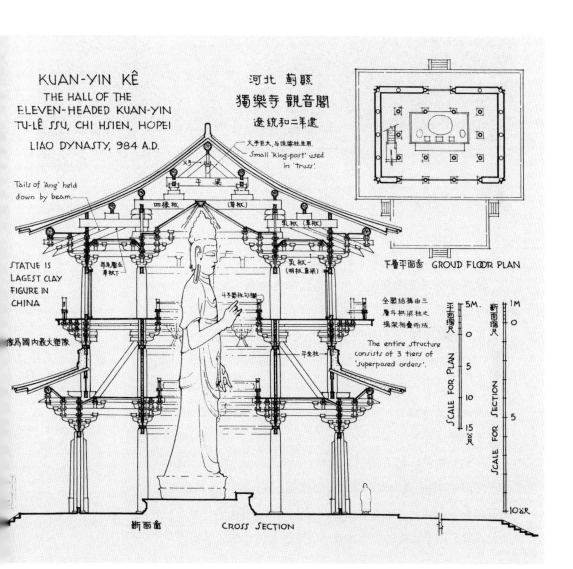

KUAN-YIN KÊ
THE HALL OF THE
ELEVEN-HEADED KUAN-YIN
TU-LÊ SSU, CHI HSIEN, HOPEI
LIAO DYNASTY, 984 A.D.

河北 薊縣
獨樂寺 觀音閣
遼統和二年建

父子巨大，与侏儒柱並用
Small 'king-post' used
in 'truss'.

父子
平梁
四椽栿
乳栿（草栿）

Tails of 'Ang' held
down by beam.

STATUE IS
LAGEST CLAY
FIGURE IN
CHINA

像為國內最大塑像

耍頭壓在
華栿下

斗子蜀柱勾欄

乳栿
（明栿月梁）

平坐柱

下層平面圖 GROUND FLOOR PLAN

全閣結構由三
層斗栱梁柱之
塌架相疊而成。

The entire structure
consists of 3 tiers of
'superposed orders'.

平面縮尺   斷面縮尺
SCALE FOR PLAN   SCALE FOR SECTION

5M.   1M
0   0

5

10   5

15

尺   10尺

斷面面   CROSS SECTION

g. 平面和断面图
g. Plan and cross
section

们所说的豪劲时期，全部位于辽代统治的华北地区。

较独乐寺观音阁和山门稍晚的辽木构建筑，是建于公元1020年［开泰九年］的辽宁义县奉国寺大殿（图27）。在这座位于关外的大殿里，其补间铺作采用和柱头铺作同样的形式，都是"双杪双下昂"，如观音阁的柱头铺作那样。其转角铺作较以往复杂，即沿角柱的两边各加了一个辅助的栌斗，也就是说，把两朵补间铺作和一朵转角铺作结合起来。这种称为附角栌斗的做法后来相当普遍，但在此早期颇属罕见（图30i）。

河北宝坻县广济寺的三大士殿，建于公元1025年［辽太平五年］，外观非常严谨，但内部却异常优雅（图28）。斗栱构造很简单，只有华栱。内部是彻上露明造，没有天花遮掩其结构特征。这种做法使匠师有一个极好的机会来表现其掌握大木作的艺术创造天才。

大同的两个建筑群

山西大同有两组建筑极为重要，即西门内的华严寺和南门内的善化寺。据记载，两寺都始建于唐代，但现存建筑物却不过是辽代

图 26
独乐寺山门
26
Main Gate（Shan
Men），Tu-le Ssu
a. 院内所见山门
全貌
a. General view
from the inner
courtyard

b. 正脊鸱吻细部，除用了唐、宋、辽各代常见的鳍形外，还有噙住正脊的龙头。鳍的上端内垂，为当时特征，此形在此后五十年内有所演变，再后则大为改观

b. Detail of ridge-end ornament. To the fin shape common in the T'ang, Sung, and Liao dynasties is added a dragon's head biting the ridge with wide-open jaws. The upper tip of the fin turns inward and down, a characteristic that was modified in fifty years and later drastically changed

c. 平面和断面图
c. Plan and cross section

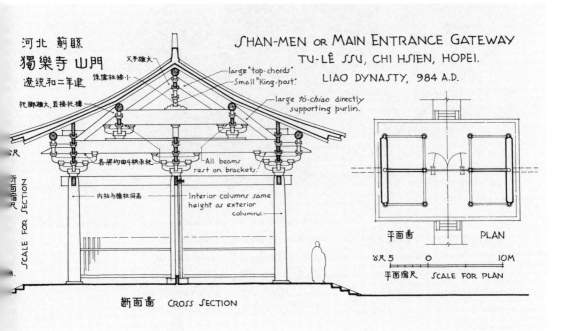

河北 蓟县
獨樂寺 山門
遼統和二年建

SHAN-MEN OR MAIN ENTRANCE GATEWAY
TU-LÊ SSU, CHI HSIEN, HOPEI.
LIAO DYNASTY, 984 A.D.

大叉手雄
小佺儒柱矮
托脚雄大，直接托槫
large "top-chords"
Small "King-post"
large tó-chiao directly supporting purlin.

各梁均由斗栱承托
All beams rest on brackets.

内柱与檐柱同高
Interior columns same height as exterior columns.

SCALE FOR SECTION

断面圖 CROSS SECTION

平面圖 PLAN

8R 5 0 10M
平面缩尺 SCALE FOR PLAN

<table>
</table>

图 27　　　　　　　a. 全景

辽宁义县奉国寺　　a. General view

大殿

　27

Main Hall, Feng-

kuo Ssu, I Hsien,

Liaoning, 1020

b. 外檐斗栱阑额上

之横木（普拍枋）

此时开始出现，但

在 金 代（ 约 1150

年）之后这种做法

已很普遍（见图 38）

b. Exterior *tou-kung*.　　ca.1150（see fig.38）

The plate above

the lintel signals

the beginning of a

practice that became

very common after

the Chin dynasty,

图 28

河北宝坻广济寺三
大士殿（已毁）

28

San-ta-shih Tien,
Main Hall, Kuang-
chi Ssu, Pao-
ti, Hopei,1025
（Destroyed）

a. 外观

a. Exterior view

b. 平面和断面图。
在这座不大的辽代
建筑中，外檐斗拱
没有斜置的下昂；
内部梁和枋都以斗
拱相交结

b. Plan and cross
section. In this
small Liao building
the exterior *tou-
kung* have no
slanting *ang*. The
ties and beams in
the interior are
assembled with *tou-
kung* at their points
of junction

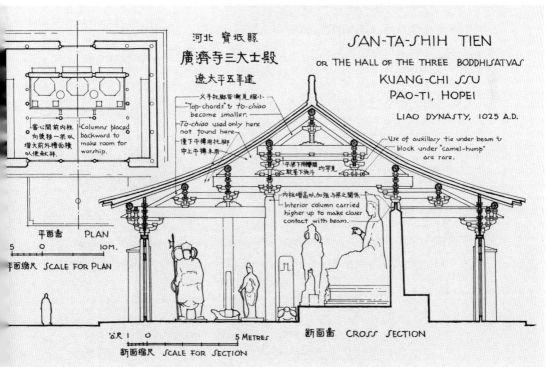

河北 寶坻縣
廣濟寺三大士殿
遼太平五年建

当心间前内柱
向後移一架以
增大前外槽面積
以便献拜
Columns placed
backward to
make room for
worship.

平面圖 PLAN

平面縮尺 SCALE FOR PLAN
5 　 10M.

义手托脚皆渐見縮小
"Top-chords" & *to-chiao*
become smaller.
To-chiao used only here
not found here —
檩下平槫用托脚
中上平槫未用

平梁下用襻間
駝峯下施斗
约罕見

内柱增高以加強
与梁之關係
Interior column carried
higher up to make closer
contact with beam.

SAN-TA-SHIH TIEN
OR THE HALL OF THE THREE BODDHISATVAS
KUANG-CHI SSU
PAO-TI, HOPEI

LIAO DYNASTY, 1025 A.D.

Use of auxillary tie under beam &
block under "camel-hump"
are rare.

斷面圖 CROSS SECTION

'6R 1　0　　　　　5 METRES
斷面縮尺 SCALE FOR SECTION

中期遗构。

华严寺原为一大寺庙，占地甚广，但在不断的边境战争中遭到很大破坏，如今仅存辽、金时代的建筑三座。其中的一座藏经的薄伽教藏殿〔图29a—图29d〕及其配殿〔海会殿，已毁于解放初期〕为辽代建筑，前者建于公元1038年〔辽重熙七年〕，后者大约也建于同一时期。〔华严寺的大殿，即今所谓"上寺"的主要建筑应属下一时期。〕

薄伽教藏殿的斗栱与观音阁相似，但内部结构为天花所遮。沿殿内两侧及后墙为藏经的壁橱，做工精致，极富建筑意味，是当时室内装修（小木作）的一个实例。其价值不仅在这里，还在于它是《营造法式》中所谓壁藏的一个实例，同时也可作为研究辽代建筑的一座极好模型。殿内还有一批出色的佛像和菩萨像。

配殿规模较小，为悬山顶。斗栱简单。值得注意的是，在栌斗中用了一根替木，作为华栱下面的一个附加的半栱。这种特别的做法只见于极少数辽代建筑，以后即不再见〔图29e〕。

华严寺建筑群的另一异常的特点是朝向。与主体建筑朝南的正统做法不同，这里的主要建筑都朝东。这是契丹人的古老习俗，他

图29

山西大同华严寺薄伽教藏殿

29

Library, Hua-yen Ssu, Tat'ung, Shansi, 1038

a. 正立面图

a. Front elevation

b. 西立面图

b. Elevation of interior wall sutra cabinet

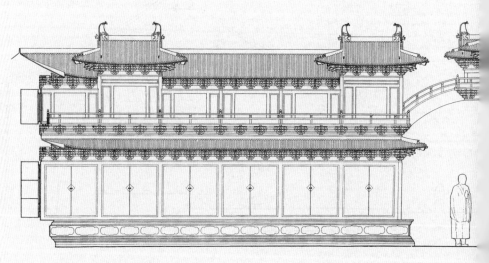

藏壁殿藏

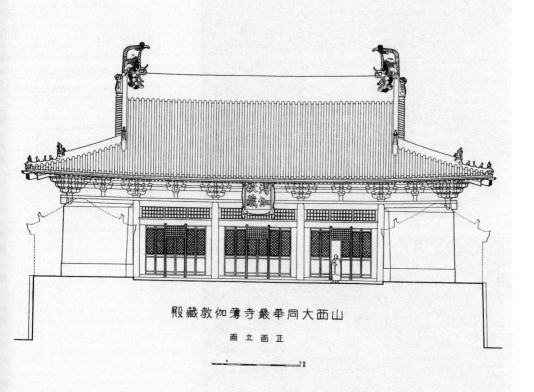

山西大同華嚴寺薄伽教藏殿

正面立面

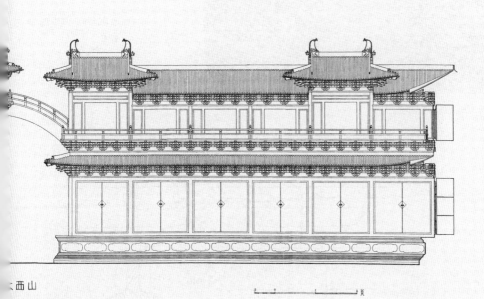

山西大

c. 壁藏圜桥细部
c. Detail of cabinetwork, arched bridge

d. 壁藏细部。这是所见最早的斜向华栱实例
d. Detail of cabinetwork, showing diagonal *hua-kung*, one of the earliest appearances of this architectural form

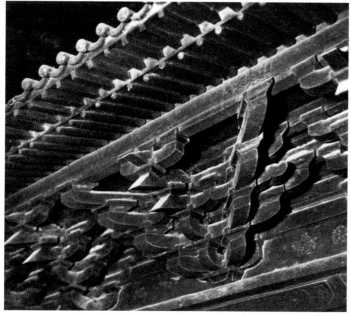

e. 配殿斗栱中的替木
e. *T'i-mu* in the Library Side Hall brackets

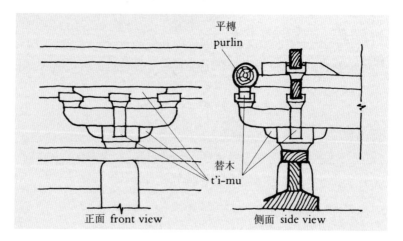

平榑
purlin

替木
t'i-mu

正面 front view 　　　　 侧面 side view

们早先崇拜太阳神，认为东是四方之首。

善化寺是大致保存了原来布局的一个建筑群（图30a，图30b）。从现存情况来看，原建筑群包括一条主轴线和两条横轴线上的七座殿。整群建筑原先四周有长廊围绕，但现在已毁，仅存基石。七座殿中只有内院一侧的一座阁被毁，其余六座仍是辽金时代原物。各殿之间的走廊及僧房已不存。

六座建筑中，大雄宝殿及普贤阁属豪劲时期，大殿（图30c—图30e）下有高阶基，广七间，两侧各有一朵殿，三殿都朝南。朵殿的设置是一种早期传统，后来已很罕见。斗栱较简单，但有一个重要特点，即在当中三间的补间铺作上使用了斜栱。这一做法曾见于华严寺薄伽教藏殿壁藏（图29d），后来在金代曾风行一时。在土墼墙内，用了横置的木骨来加固，有效地防止了竖向开裂。这种办法还见于13—14世纪的某些建筑，但并不普遍。

普贤阁（图30f—图30h）有两层，规模很小，结构上与独乐寺观音阁基本相同，也使用了斜栱。

以上两座建筑为辽代遗物，确切年代已不可考，但从形制特征

图30
山西大同善化寺。
建于辽金时期［约
公元1060年］
30
Shan-hua Ssu, Ta-
t'ung, Shansi, Liao
and Chin
a. 全景
a. General view

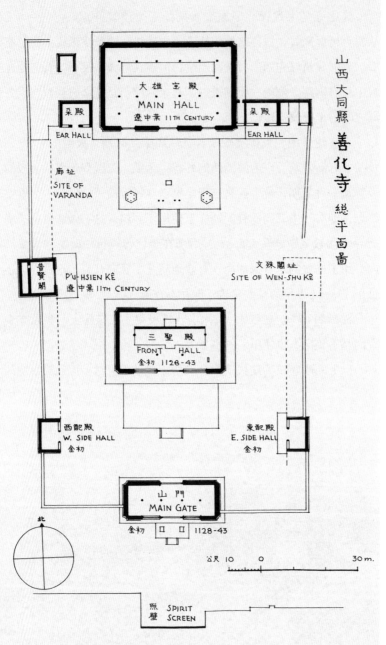

山西大同縣 善化寺 總平面圖

大雄宝殿
MAIN HALL
遼中葉 11TH CENTURY

朵殿
EAR HALL

朵殿
EAR HALL

廊址
SITE OF VARANDA

普賢閣
P'u-HSIEN KÊ
遼中葉 11TH CENTURY

文殊閣址
SITE OF WEN-SHU KÊ

三聖殿
FRONT HALL
金初 1128-43

西配殿
W. SIDE HALL
金初

東配殿
E. SIDE HALL
金初

北

山門
MAIN GATE
金初 1128-43

公尺 10   0   30 m.

照壁
SPIRIT SCREEN

· PLOT PLAN · SHAN-HUA SSU · TA-T'UNG · SHANSI ·

b. 总平面图。这
是大多数中国佛
教、道教寺观的典
型布局。大殿位于
中轴线上，较小的
殿和配殿则在横轴
线上。各殿以廊相
接，形成一进进的
长方形庭院

b. Site plan. This
plan is typical
of most Chinese
temples, Buddhist
or Taoist. The
main hall or halls
are placed on the
central axis, minor
halls or subsidiary
buildings on
transverse axes.
The buildings are
usually connected
by galleries and
form a series
of rectangular
courtyards

c. 大殿立面渲染图
c. Main Hall, ca. 1060, rendering
d. 大殿内梁架及斗栱
d. Main Hall, roof frame

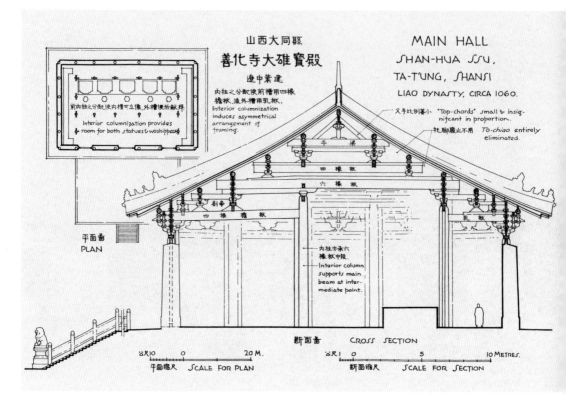

山西大同縣
善化寺大雄寶殿
遼中葉建

MAIN HALL
SHAN-HUA SSU,
TA-T'UNG, SHANSI

LIAO DYNASTY, CIRCA 1060.

内柱之分配使前槽用四椽
栿栿,後外槽使用乳栿.
Interior columnization
induces asymmetrical
arrangement of
framing.

叉手比例較小 "Top-chords" small & insig-
nifcant in proportion.

托脚叢止不用 Tō-chiao entirely
eliminated.

前内柱之分配使内槽可立像,外槽使外敬拜
interior columnization provides
& room for both statues & worshippers.

平樑

四椽栿

六椽栿

劄牽

四椽栿

乳栿

平面圖
PLAN

内柱承六
椽栿中段
Interior column
supports main
beam at inter-
mediate point.

断面圖　CROSS SECTION

古R10　0　20 M.
平面縮尺　SCALE FOR PLAN

古R1 0　5　10 METRES.
断面縮尺　SCALE FOR SECTION

e. 大殿平面和断面图

e. Main Hall, plan and cross section

f. 善化寺普贤阁

f. P'u-hsien Ke(Hall of Samantabhadra）, ca. 1060

g. 普贤阁立面渲染图

g. P'u-hsien Ke, rendering

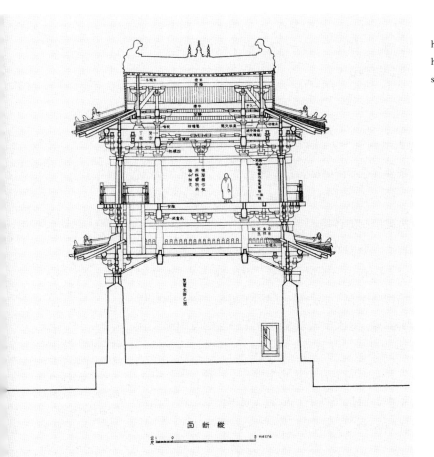

山西大同善化寺普賢閣

縱斷面

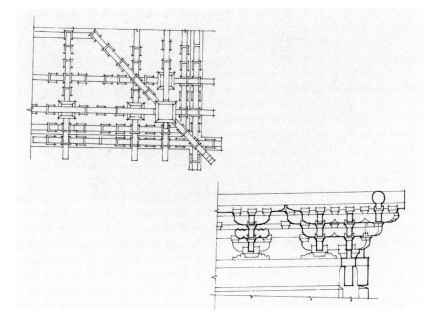

i. 转角铺作加附角栌斗的平面和立面图
i. Plan and section of a *fu-chiao lu-tou* （corner set with adjoining *lu-tou*）
［h、i 两图为后来新加，不在英文原稿之内］

上看似应属 11 世纪中叶。善化寺前殿［三圣殿］及山门属于下一时期，我们将在下文加以讨论。

佛宫寺木塔

山西应县佛宫寺［释迦］木塔（图 31）可被看作豪劲时期建筑的一个辉煌尾声。塔建于公元 1056 年［辽清宁二年］，可能是该时代的一种常见形制，因为在当年属辽统治地区的河北、热河、辽宁诸省还可见到少数仿此形制的砖塔。

塔的平面为八角形，有内外两周柱，五层全部木构。其结构的基本原则与独乐寺观音阁相近：除第一层外，其上四层之下都有平坐，实际上是九层结构叠架在一起。第一层周围檐柱之外，更加以单坡屋顶［“周匝副阶”］，造成重檐效果。最高一层的八角攒尖顶冠以铁刹，自地平至刹端高 183 英尺［原文有误，应为 220 英尺，约 67 米］，整座建筑共有不同组合形式的斗栱五十六种，我们在上文中提到过的所有各种都包括在内了。对于研究中国建筑的学生来说，这真是一套最好的标本。

c. 渲染图

c. Rendering

山西應縣佛宮寺遼釋迦木塔

d. 断面图。叠置的各层檐柱逐层内缩，使全塔略呈锥形，但除最高一层外，其全部内柱都置于同一垂直线上。最下面两层的外檐由双杪双下昂的斗栱承托，上面三层及所有平坐的斗栱出跳则只用华栱（右页）

d. Section. The exterior superposed orders are set back slightly on each floor so that the building tapers from base to top floor the inside columns are carried up in a continuous line. The eaves of the two lower stories are supported by *tou-kung* and two tiers of *ang*; the three upper stories and all the balconies use *hua-kung* only

東西断面

山西應縣佛宮寺遼釋迦木塔

中國營造學社測繪　民國廿三年九月實測　廿四年六月製圖

醇和时期

（约公元 1000—1400 年）

宋初建筑的特征

当辽代的契丹人尚在恪守唐代严谨遗风的时候，宋朝的建筑家却已创出了一种以典雅优美为其特征的新格调。这一时期前后延续了约四百年。

在风格的演变中最引人注目的，就是斗栱规格的逐渐缩小，到公元 1400 年前后已从柱高的约三分之一缩至约四分之一（图 32）。补间铺作的规格却相对地越来越大，组合越来越复杂，最后，不仅其形制已和柱头铺作完全相同，而且由于采用了斜栱，其复杂程度甚至有过之而无不及。这种补间铺作，由于上不承梁，下不落柱而增加了阑额的负担。据《营造法式》及某些实例，当心间用两朵补间铺作。在补间铺作用下昂的情况下，由于昂尾向上斜伸，使结构问题更加复杂。在柱头铺作中，昂尾压在梁头之下，以固定其位置；而在补间铺作中，却巧妙地成为上面的槫在梁架之间的支托。昂的设置使建筑家有了一个施展才华的大好机会，得以构造出种种不同的、极其有趣的铺作，但其结构功能从未被忽视。铺作的设置总是在支承整座建筑中各司其职，罕有不起作用或纯粹为了装饰的。

在建筑内部，柱的布置方式常根据建筑物的用途而加以调整，如留出地位来放置佛像、容纳朝拜者等等。当内柱减少时，则不仅平面布置，梁架也会受到影响（下文将讨论这种情况的几个实例）。除柱的这种不规则布置以外，常加高内柱，以直接支承上面的梁。但在任何两个横竖构件相交的地方，总要用一组简单的斗栱

图 32
历代斗栱演变图

32
Evolution of the
Chinese "order"

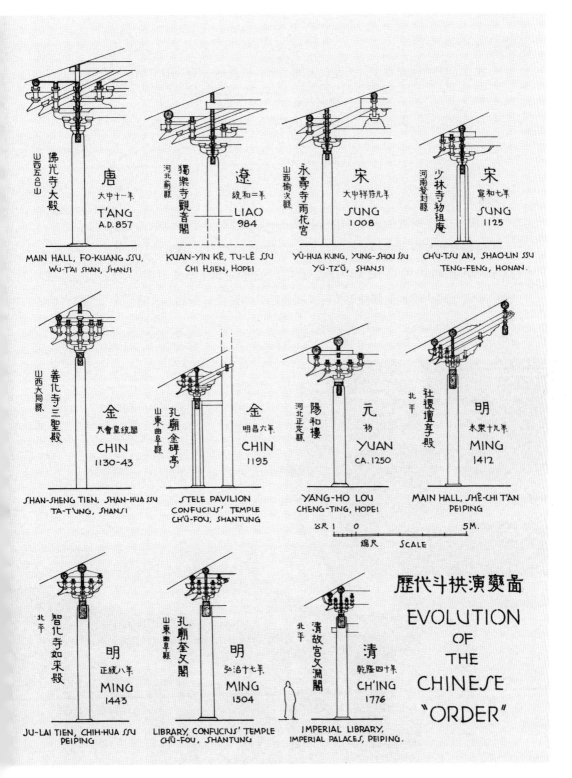

佛光寺大殿
山西五台山
唐
大中十一年
T'ANG
A.D. 857

MAIN HALL, FO-KUANG SSU,
WU-T'AI SHAN, SHANSI

獨樂寺觀音閣
河北薊縣
遼
統和二年
LIAO
984

KUAN-YIN KÊ, TU-LÊ SSU
CHI HSIEN, HOPEI

永壽寺雨花宮
山西榆次縣
宋
大中祥符元年
SUNG
1008

YÜ-HUA KUNG, YUNG-SHOU SSU
YÜ-TZ'Ü, SHANSI

少林寺初祖庵
河南登封縣
宋
宣和七年
SUNG
1125

CH'U-TSU AN, SHAO-LIN SSU
TENG-FENG, HONAN.

善化寺三聖殿
山西大同縣
金
天會皇統間
CHIN
1130-43

SHAN-SHENG TIEN, SHAN-HUA SSU
TA-T'UNG, SHANSI

孔廟金碑亭
山東曲阜縣
金
明昌六年
CHIN
1195

STELE PAVILION
CONFUCIUS' TEMPLE
CH'Ü-FOU, SHANTUNG

陽和樓
河北正定縣
元
初
YUAN
CA. 1250

YANG-HO LOU
CHENG-TING, HOPEI

社稷壇享殿
北平
明
永樂十九年
MING
1412

MAIN HALL, SHÊ-CHI T'AN
PEIPING

營造尺 1   0   5 M.
縮尺   SCALE

智化寺如來殿
北平
明
正統八年
MING
1443

JU-LAI TIEN, CHIH-HUA SSU
PEIPING

孔廟奎文閣
山東曲阜縣
明
弘治十七年
MING
1504

LIBRARY, CONFUCIUS' TEMPLE
CHÜ-FOU, SHANTUNG

清故宮文淵閣
北平
清
乾隆四十年
CH'ING
1776

IMPERIAL LIBRARY,
IMPERIAL PALACES, PEIPING.

歷代斗栱演變圖
**EVOLUTION
OF
THE
CHINESE
"ORDER"**

77

来过渡。

　　在有天花的建筑物中，天花以上的梁架各构件的表面一般不加工或刨整。但在彻上露明造的殿堂中，却显示出各种梁枋、斗栱、昂尾等错综复杂、彼此倚靠的情形。其配置的艺术正是这一时期匠师之所长。

一处先驱者：雨华宫

　　山西榆次县附近的永寿寺中的小殿——雨华宫，是一座以熟练的手法将醇和与豪劲两种风格融合为一的建筑物（图 33）。按其年代之早（公元 1008 年）[宋大中祥符元年]，它本应属于前一时期。但后来宋元时代建筑的那种柔美特征在此已见端倪。它是集两个时期的特征于一身的一个过渡型实例。

b. 外廊上部构架
b. Framing over porch

a. 全景
a. General view

c. 永寿寺雨华宫半立面和半纵断面图

c. Elevation and longitudinal section

d. 永寿寺雨华宫平面和断面图

d. Plan and cross section

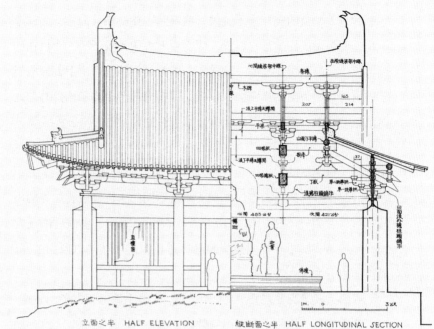

立面之半 HALF ELEVATION　　　　纵断面之半 HALF LONGITUDINAL SECTION

山西榆次縣　永壽寺雨華宮　宋大中祥符元年建

YÜ-HUA KUNG · MAIN HALL OF YUNG-SHOU SSU
YÜ-TZ'U · SHANSI · 1008 A·D·

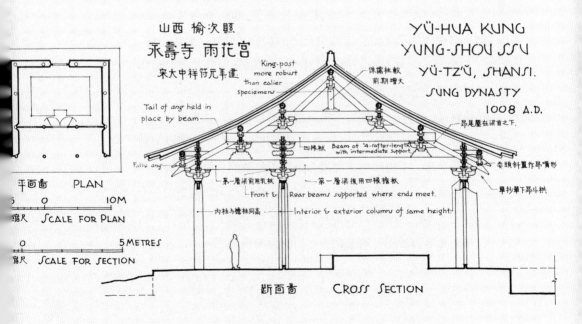

山西 榆次縣
永壽寺 雨花宮
宋大中祥符元年建

YÜ-HUA KUNG
YUNG-SHOU SSU
YÜ-TZ'Ŭ, SHANSI.
SUNG DYNASTY
1008 A.D.

King-post more robust than earlier Speciemens

Tail of ang held in place by beam

休儒柱较前期增大

昂尾壓在梁首之下.

委頭斜置作昂"嘴形

四椽栿　Beam of "4-rafter-length" with intermediate support.

False ang

第一層梁前用乳栿　第一層梁後用四椽檐栿

Front & Rear beams supported where ends meet.

內柱与檐柱同高　Interior & exterior columns of same height.

單抄單下昂斗拱

平面圖　PLAN

縮尺　SCALE FOR PLAN

5　0　10M

0　5 METRES

節尺　SCALE FOR SECTION

斷面圖　CROSS SECTION

这座建筑既不宏伟，又已破败，因而乍看起来并不吸引人。但它那种令人愉快的美却逃不过内行的眼睛。斗栱极其简单，单杪单下昂，要头做成昂嘴形，斜置，使之看上去像是双下昂。斗栱在比例上较大，略小于柱高的三分之一，因此补间铺作事实上就被取消了。内部梁架为彻上露明造，包括昂尾在内的各个露明构件都如此简洁地结合在一起，看得出是遵循着严密逻辑而得出的必然结果。

正定的一组建筑

河北正定县的隆兴寺，保存了一批早期的宋代建筑物。寺的山门尽管保存得还不错，却是 18 世纪（清乾隆时期）重修后的混合物，一些清式的小斗栱竟被生硬地塞进巨大的宋代斗栱原物之间，显得不伦不类。

寺的大殿名为摩尼殿（图 34a，图 34b），殿平面近正方形，重檐。四面各出抱厦，抱厦屋顶以山墙朝向正面［"出际"向前］。这种做法常可见于古代绘画，但实物却很难得。斗栱大而敦实，虽然每间只用补间铺作一朵，但有辽代惯用的斜栱。檐柱明显地向屋角渐次加高，给人一种和缓感。

转轮藏殿是一座为了安置转轮藏而建造的殿（图 34c—图 34g）。殿中对内柱的位置做了改动，为转轮藏让出了空间。而这又影响到上层彻上露明造的梁架结构，其中众多的构件巧妙地结合为一体，犹如一首演奏得极好的交响曲，其中每个乐部都准确而及时地出现，真正达到了完美、和谐的境地。

转轮藏是一个中有立轴的八角形旋转书架，为此类构造中一个罕见的实例。它的外形如一座重檐亭子，建筑构件的处理极为精致。下檐八角形，上檐圆形，两檐都采用了复杂的斗栱。由于这项小木作严格遵循了《营造法式》中的规定，所以是宋代构造的一个极有价值的实例。遗憾的是，当笔者于 1933 年最后一次见到时，该寺正被当作兵营使用，而它在士兵们野蛮的糟害之下，已经破败不堪了。

图 34

河北正定隆兴寺

34

Lung-hsing Ssu,
Cheng-ting, Hopei

a. 摩尼殿平面图

a. Plan of Mo-ni Tien

b. 摩尼殿，建于宋
初，约公元1030
（？）年（1978年该
殿大修时多处发
现墨迹题记，证
明系建于北宋皇
祐四年，即公元
1052年）

b. Mo-ni Tien, Main
Hall, early Sung, ca.
1030（？）

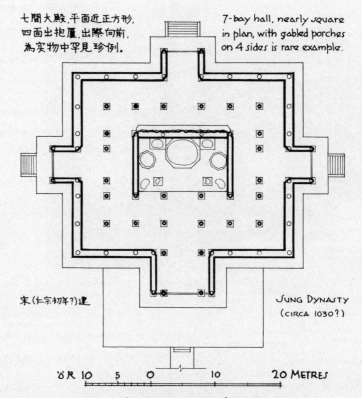

七間大殿，平面近正方形，四面出抱厦，出際向前，為实物中罕見珍例。

7-bay hall, nearly square in plan, with gabled porches on 4 sides is rare example.

宋（仁宗初年?)建

JUNG DYNASTY (CIRCA 1030?)

0尺 10 5 0 10 20 METRES

河北正定縣龍興寺摩尼殿平面

PLAN, MO-NI TIEN, LUNG-HSING SSU, CHEN-TING.

c. 转轮藏殿，建
于宋初，约公元
960—1126 年
c. Library, Chuan-
lun-tsang Tien
（The Hall of
the Revolving
Bookcase）,early
Sung, ca.960-1126

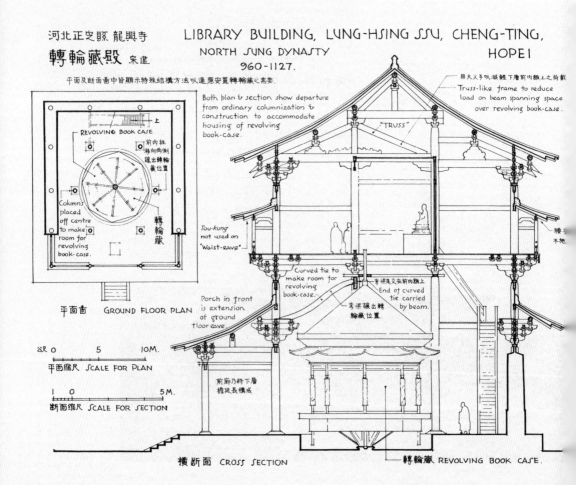

河北正定縣 龍興寺
轉輪藏殿 宋建

LIBRARY BUILDING, LUNG-HSING SSU, CHENG-TING,
NORTH SUNG DYNASTY   HOPEI
960-1127.

平面及斷面畫中皆顯示特殊結構方法以適應安置轉輪藏之需要.

Both plan & section show departure
from ordinary columnization &
construction to accommodate
housing of revolving book-case.

REVOLVING BOOK CASE

上

前内柱
移向兩側
讓出轉輪
藏位置

轉輪藏

Columns
placed
off centre
to make
room for
revolving
book-case.

平面圖 GROUND FLOOR PLAN

Porch in front
is extension
of ground
floor eave.

前廊乃將下層
檐延長構成

比例 0   5   10M.
平面縮尺 SCALE FOR PLAN

1   0   5M.
斷面縮尺 SCALE FOR SECTION

用大义手以減輕下層前内額上之荷載
Truss-like frame to reduce
load on beam spanning space
over revolving book-case.

"TRUSS"

Tou-kung
not used on
"Waist-eave"

腰檐
不施

Curved tie to
make room for
revolving
book-case.

弯梁足叉在前内額上
End of curved
tie carried
by beam.

弯梁讓出轉
輪藏位置

橫斷面 CROSS SECTION

轉輪藏 REVOLVING BOOK CASE.

e. 转轮藏殿屋顶下结构。注意其对结构构件的具有特色的处理

e. Library, the large truss under the floor. Note the characteristic decorative treatment of the structural members

f. 转轮藏殿内部转角铺作

f. Library, interior corner *tou-kung*

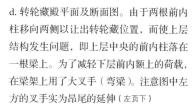

d. 转轮藏殿平面及断面图。由于两根前内柱移向两侧以让出转轮藏位置，而使上层结构发生问题，即上层中央的前内柱落在一根梁上。为了减轻下层前内额上的荷载，在梁架上用了大叉手（弯梁）。注意图中左方的叉手实为昂尾的延伸（左页下）

d. Library, plan and cross section. Two interior columns are place out of line to accommodate the revolving cabinet. This irregularity created a structural problem for the floor above, where the central column stands on a girder. The solution was the introduction of a large truss with chords to divert the load of the beam above to the columns at front and rear. Note that the left chord is the extension of the tail of the *ang*

g. 转轮藏

g. Library, the revolving sutra cabinet

83

晋祠建筑群

　　山西太原近郊晋祠中的圣母殿（图35），是另一组重要的宋代早期建筑。包括一座重檐正殿；殿前为一座桥［飞梁］，桥下是一个长方形的水池［鱼沼］；再往前是一座献殿和一座牌楼；再前是一个平台，上有四尊铁铸太尉像［金人台］。桥和两座殿都建于宋天圣年间（公元 1023—1031 年）。除彩画外，全组建筑保存完好，虽经历代修缮，却基本未损原貌。

　　隆兴寺和晋祠建筑的斗栱在其昂嘴形制上还有一个突出的特色（图36）。在这以前的建筑中，昂嘴都是一个简单的斜面，截面呈长方形。斜面与底面形成约 25 度夹角。这在《营造法式》中被称为"批竹昂"。但在这两组建筑中，斜面部分的上方却略为隆起，使其横截面上部呈半圆形，称为"琴面昂"。第三种做法是使斜面下凹，但横截面上部仍然隆起，其形状犹如一管状圆环之内侧。自《营造法式》时代直到后来，这第三种做法已成定制，只是横截面上部的隆起已简化为将斜面两侧棱角削去而已。在这两组建筑中所有昂都取第二种做法。这种做法只见于 11 世纪早期，或许还见于 10 世纪晚期的宋代建筑中。因为自《营造法式》刊布（公元 1103 年）之

図 35
山西太原近郊晋祠圣母殿，建于宋代，约公元 1030 年
35
Sheng Mu Miao（Temple of the Saintly Mother），Chin-tz'u, near T'aiyuan, Shansi, early Sung, ca.1030

a. 大殿全景
a. Main Hall, general view

b. 大殿正面细部,
图中可见斗栱及昂
b. Main Hall, detail
of facade, showing
tou-kung and *ang*

c. 大殿前廊内景
c. Main Hall,
interior of porch

d. 献殿
d. Front Hall

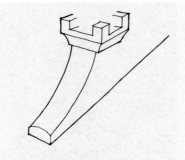

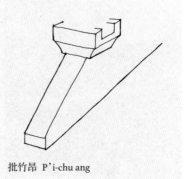

第三种做法的昂（原文误为"琴面昂"）Ch'ing-mien ang

批竹昂 P'i-chu ang

图 36
昂嘴的演变
36
Changes in shape of
beak of the *ang*

后，第三种做法已成为标准形制。我们只在隆兴寺的摩尼殿和转轮藏殿以及佛光寺的文殊殿中见到过第二种做法，这表明它们大体上属于同一时期。[此段英文原文有误，图 36 的图注亦有误。现根据梁思成原始手稿订正译出。图 36 右为批竹昂，左应是第三种做法的昂，不是文中所指的琴面昂。]

这一短时期建筑的另一特点，是将耍头做成与昂嘴完全一样的形状（图 37）。在早期建筑中，耍头或是方形的，或是一个简单的批竹形；后来则做成夔龙头形或蚂蚱头形。把耍头完全做成昂状，并与昂取同一角度，使人产生双下昂的印象，这种做法仅见于 11 世纪的建筑物。

谈到这里，我们还应提到普拍枋。这是直接置于阑额之上的一根横木（图 38）。它与阑额相叠，形成 T 字形的截面。这种做法最早见于山西榆次的雨华宫，在正定和晋祠的建筑中也可见到。但在早期建筑中，这种做法比较少见。即使按《营造法式》规定，这种做法也只用于平坐。但到了金、南宋以后（约公元 1150 年），普拍枋已普遍用于所有建筑，阑额上不置枋者反倒成了例外。

晋祠建筑的斗栱上另一个新东西是所谓假昂，即把平置的华栱的外端斫作昂嘴形状。这样，虽然没有采用斜置的下昂，却获得了昂嘴的装饰效果。华栱成了一个附加装饰品，这是一种退化的标

图 37
历代耍头（梁头）
演变图（右页）
37
Evolution of the
shua t'ou（head of
the beam）

图像中国建筑史

歷代耍頭(梁頭)演變圖 EVOLUTION OF THE S'HUA-T'OU (HEAD OF THE BEAM)

公分 10 0   50   100 cm.

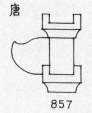

唐

857

佛光寺正殿

MAIN HALL, FO-KUANG SSU

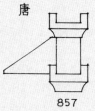

唐

857

佛光寺正殿

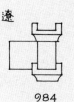

遼

984

獨樂寺觀音閣

TU-LÊ SSU

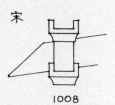

宋

1008

永壽寺雨華宮

YUNG-SHOU SSU

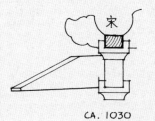

宋

CA. 1030

佛光寺文殊殿

WEN-SHU TIEN, FO-KUANG SSU

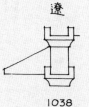

遼

1038

薄伽教藏

LIBRARY HUA-YEN SSU

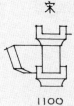

宋

1100

營造法式

YING-TSAO FA-SHIH

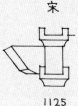

宋

1125

初祖庵

CH'U-TSU AN

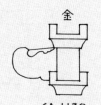

金

CA. 1130

華嚴寺大殿

MAIN HALL, HUA-YEN SSU

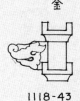

金

1118-43

善化寺三聖殿

FRONT HALL SHAN-HUA SSU

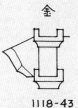

金

1118-43

善化寺三聖殿

FRONT HALL SHAN-HUA SSU

金

1118-43

善化寺山門

MAIN GATE SHAN-HUA SSU

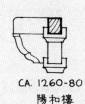

元

CA. 1260-80

陽和樓

YANG-HO LOU

明

1504

奎文閣

LIBRARY CONFUCIUS' TEMPLE

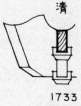

清

1733

工程做法

KUNG-CH'ENG TSO-FA CHÊ-LI

清

1776

文淵閣

WEN-YUAN KÊ

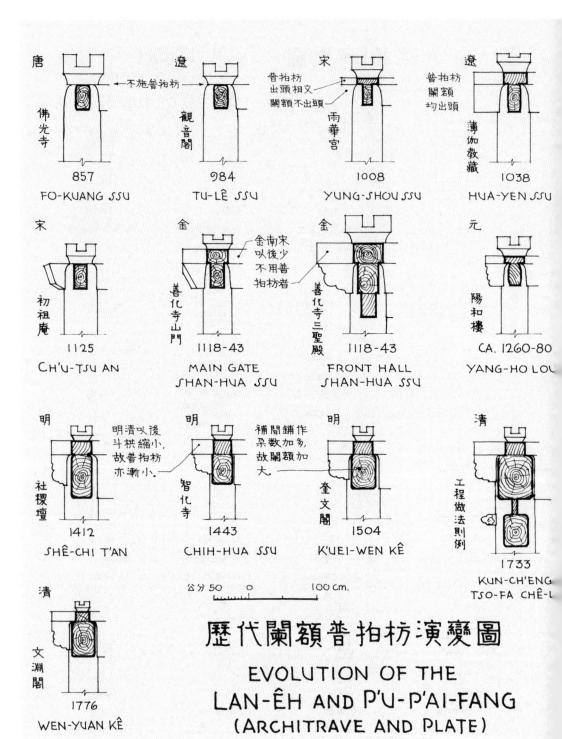

歷代闌額普拍枋演變圖

EVOLUTION OF THE
LAN-ÊH AND P'U-P'AI-FANG
(ARCHITRAVE AND PLATE)

唐　佛光寺　857　FO-KUANG SSU

遼　觀音閣　984　TU-LÊ SSU

宋　不施普拍枋　雨華宮　普拍枋出頭相文　闌額不出頭　1008　YUNG-SHOU SSU

遼　普拍枋闌額均出頭　薄伽教藏　1038　HUA-YEN SSU

宋　初祖庵　1125　CH'U-TSU AN

金　善化寺山門　金南宋以後少不用普拍枋者　1118-43　MAIN GATE SHAN-HUA SSU

金　善化寺三聖殿　1118-43　FRONT HALL SHAN-HUA SSU

元　陽和樓　CA. 1260-80　YANG-HO LOU

明　社稷壇　明清以後斗栱縮小,故普拍枋亦漸小.　1412　SHÊ-CHI T'AN

明　智化寺　1443　CHIH-HUA SSU

明　奎文閣　補間鋪作朵數加多,故闌額加大.　1504　K'UEI-WEN KÊ

清　工程做法則例　1733　KUN-CH'ENG TSO-FA CHÊ-L

清　文淵閣　1776　WEN-YUAN KÊ

公分 50　0　100 cm.

88

图 38
历代阑额、普拍枋
演变图（左页）
38
Evolution of the
lan-e and *p'u-p'ai fang*

志。这种做法最终成了明、清的定制。然而，在这样早的时期见到它，确是一种不祥之兆，预示着后来在结构上脱离朴实性的趋向。

一座独特的建筑——文殊殿

佛光寺唐代大殿的配殿文殊殿，是一栋面广七间悬山顶的殿，貌不惊人（图 39）。其斗栱做法与隆兴寺、晋祠相似。然而，其内部

图 39
佛光寺文殊殿
39
Wen-shu Tien（Hall
of Manjusr），Fo-
kuang Ssu
a. 全景
a. General view

b. 内景
b. Interior, showing
"queen-post" truss

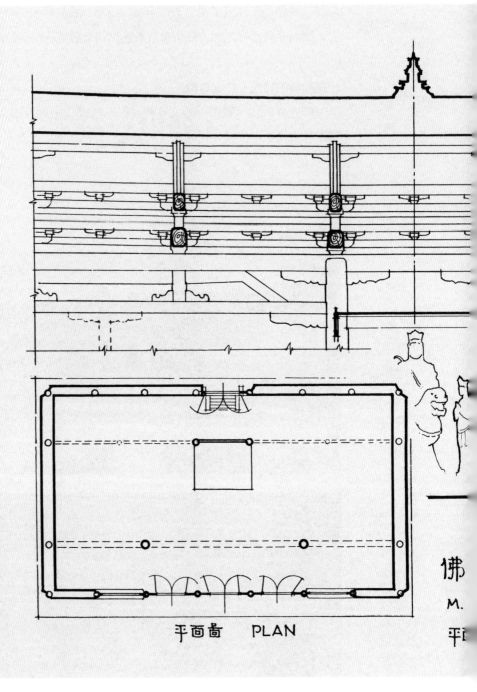

平面畫　PLAN

佛
M.
平

c. 平面及纵断面图

c. Plan and longitudinal section

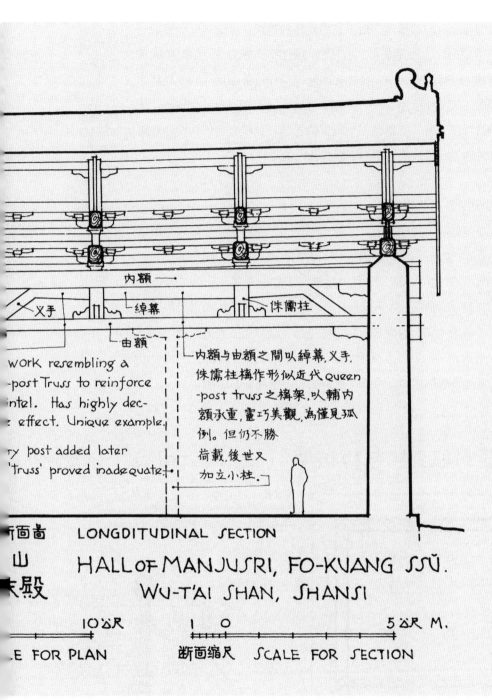

内额 →

义手　　　綽幕　　　侏儒柱

由额

work resembling a
-post Truss to reinforce
ntel. Has highly dec-
e effect. Unique example.

ry post added later
'truss' proved inadequate.

内额与由额之間以綽幕,义手,
侏儒柱欘作形似近代 Queen
-post truss 之構架,以輔内
額承重,靈巧美觀,為僅見孤
例。但仍不勝
荷載,後世又
加立小柱。

LONGDITUDINAL SECTION

HALL OF MANJUSRI, FO-KUANG SSŬ.
WU-T'AI SHAN, SHANSI

10 公尺
LE FOR PLAN

1 0
断面缩尺　SCALE FOR SECTION

5 公尺 M.

构架却是个有趣的孤例。由于它那特殊的构架，其后部仅在当心间用了两根内柱，致使其左、右柱的间距横跨三间，长度竟达 46 英尺［约 14 米］，中到中。这样大的跨度是任何普通尺寸的木料都达不到的，于是便采用了一个类似于现代双柱式桁架的复合构架。从结构的角度来看，它不是一个真正的桁架，并没有起到其设计者所预期的作用，以致后世不得不加立辅助的支柱。

《营造法式》时代遗例

在现存的宋代建筑中，其建造年代与《营造法式》最接近的是一座很小的殿——河南嵩山少林寺的初祖庵（图40）。殿为方形，三间。其石柱为八角形，其中一柱刻有宋宣和七年（公元 1125 年）字样，距《营造法式》刊布仅二十二年。其总体结构相当严格地依照了《营造法式》的规定，其斗栱更是完全遵循了有关则例。有一个次要的地方也反映了宋代建筑的特征，即踏道侧面的三角形象眼，其厚度逐层递减，恰如《营造法式》所规定的那样。

图 40
河南登封县少林寺
初祖庵
　40
Ch'u-tsu An, Shao-
lin Ssu, Sung Shan,
Teng-feng, Honan,
1125

c. 平面图及补间铺
作、柱头铺作图
c. Plan, and
intermediate and
column *tou-kung*

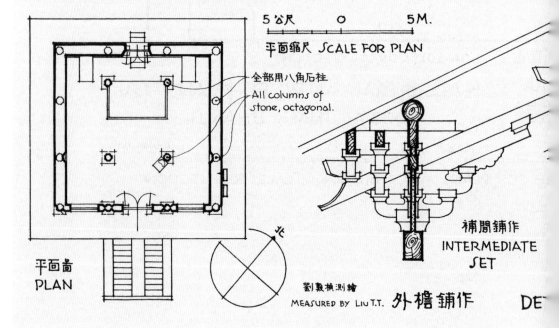

a. 全景

a. General view

b. 正面细部

b. Detail of facade

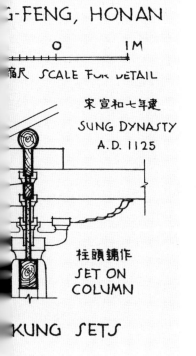

G-FENG, HONAN

0   1M

縮尺 SCALE FOR DETAIL

宋 宣和七年建

SUNG DYNASTY

A.D. 1125

柱頭鋪作

SET ON
COLUMN

KUNG SETS

由此往北，为当时金人统治地区，其中也有几座约略建于这一时期的建筑。如山西应县净土寺大殿（图41），建于公元1124年［金天会二年］，距《营造法式》的刊行时间比初祖庵还要近一年。尽管有政治上和地理上的阻隔，这座建筑的整体比例却相当严格地遵守了宋代的规定；其藻井采取了《营造法式》中天宫楼阁的做法，是一件了不起的小木作装修技术杰作。

山西大同善化寺的三圣殿和山门（图42，图43）建于公元1128—1143年［金天会六年至皇统三年间］，也在当时金人统治地区。三圣殿的补间铺作也和较早的隆兴寺摩尼殿和应县木塔等处一样，使用了斜栱，但已演变为一堆错综复杂的庞然大物，成了压在阑额上的一个沉重负担。阑额上的普拍枋也较早先的几例更厚。殿的内部是彻上露明造，其斗栱的使用颇具实效。

善化寺山门在同类建筑中可能是最为夸张的一个。这个五间的山门显然比某些小寺的正殿还要大。其斗栱较简单，未用斜栱。这里使用了当时北方已很少见的月梁。

南宋王朝与金同时。目前所知，在南方这一时期的木构建筑唯一留存至今的，是江苏苏州道教建筑玄妙观的三清殿。[1] 虽然它建于公元1179年［南宋淳熙六年］，距《营造法式》刊出年代不过七十来年，却已把当时那种豪劲风格丧失了许多。与和它同时甚至更晚的北方建筑比起来，它显得过分雕琢，特别值得注意的是斗栱与整座建筑的比例变小了。

醇和时期的最后阶段

在北方和南方，都有相当一批建于醇和时期最后一百五十年间的建筑实例留存了下来。在此期间，斗栱有了许多重要的变化：其一是假昂的普遍使用，其最早实例见于晋祠；另一是要头，即柱头铺作上梁的外端（蚂蚱头）的增大（图37）。由于斗栱随着时代的演变而缩小，依旧制应相当于一材的要头在结构上就显得过于脆弱了。因此，从比例上说，这时期的要头就必须大于一材。而为了承受它，

图41
山西应县净土寺大殿

41

Main Hall, Ching-t'u Ssu, Ying Hsien, Shansi, 1124

［1］
解放后发现尚有福州华林寺大殿，建于五代钱弘俶十八年（公元964年）及余姚保国寺大殿，建于宋大中祥符六年（公元1013年）。——孙增蕃校注

a. 正面
a. Facade

b. 藻井
b. Ceiling

图 42

山西大同善化寺三
圣殿

42

Front Hall, Shan-
hua Ssu, Ta-t'ung,
Shansi, 1128-1143

a. 全景

a. General view

b. 正面细部，其中
可见补间铺作中的
斜栱

b. Detail of facade,
showing diagonal
kung in intermediate
bracket sets

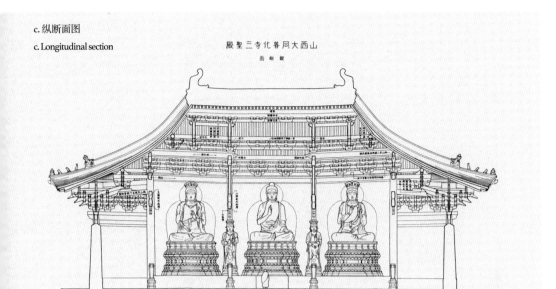

山西大同善化寺三圣殿
纵断面

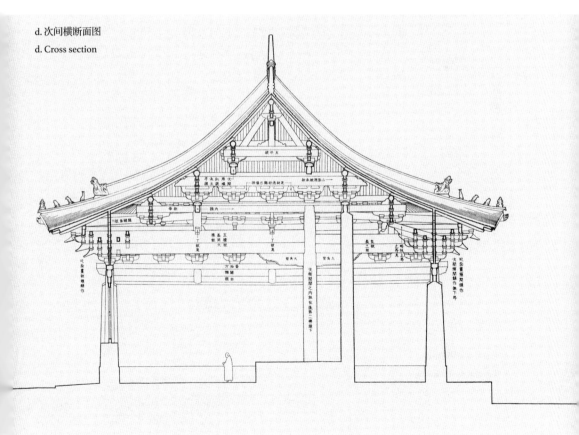

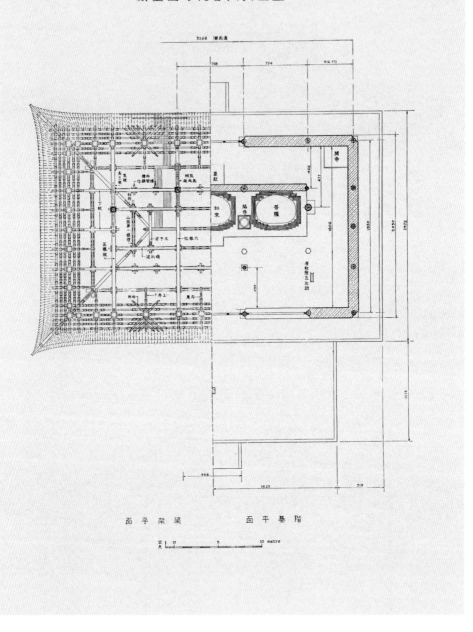

山西大同善化寺三聖殿

梁架平面　　階基平面

e. 平面及梁架平面图

e. Plan and roof framing

98

图 43

善化寺山门

43

Shan-men，Main
Gate，Shan-hua
Ssu, 1128-1143

a. 全景

a. General view

b. 平面及断面图

b. Plan and cross
section

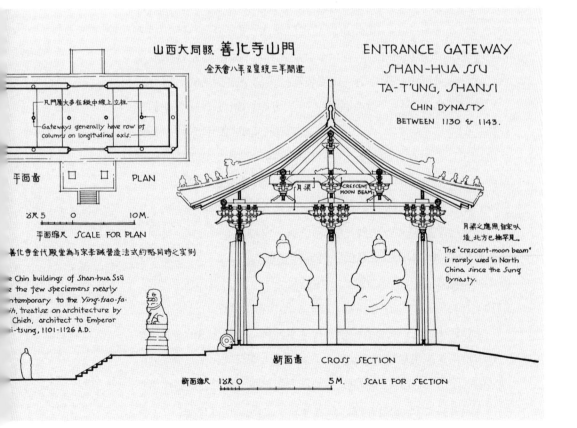

山西大同縣 善化寺山門

金天會八年至皇統三年間建

ENTRANCE GATEWAY
SHAN-HUA SSU
TA-T'UNG, SHANSI
CHIN DYNASTY
BETWEEN 1130 & 1143.

凡門屋大多在縱中線上立柱

Gateways generally have row of
columns on longitudinal axis.

平面圖   PLAN

尺5 0   10 M.

平面縮尺 SCALE FOR PLAN

善化寺金代殿堂為與宋李誠營造法式約略同時之實例

e Chin buildings of Shan-hua Ssŭ
e the few speciemens nearly
ntemporary to the Ying-tsao-fa-
h, treatise on architecture by
Chieh, architect to Emperor
i-tsung, 1101-1126 A.D.

月梁   CRESCENT
MOON BEAM

月梁之應用,自宋以
後,北方已極罕見。

The "crescent-moon beam"
is rarely used in North
China since the Sung
Dynasty.

断面圖 CROSS SECTION

断面縮尺 尺 0   5 M. SCALE FOR SECTION

99

下面的华栱也得相应加宽（到了《工程做法则例》出版的时代，即公元 1734 年［清雍正十二年］，柱头铺作上耍头的宽度已扩大到 40 分°，即四斗口，较之宋代的比例要大四倍。而最下一跳华栱的宽度也比旧制增大了一倍，即从 10 分° 增至 20 分°，即二斗口）。

河北正定县阳和楼（约公元 1250 年）［金末或元初］，是醇和末期建筑的一个极好实例。这是一座七间的类似望楼的建筑，建于一座很高的砖台上，台下有两条发券门洞，形若城门。楼台位于城内主要大道上，像是某种纪念性建筑物（图 44）。其斗栱看去有如双下昂，实际上柱头铺作两跳昂都是假的，补间铺作的昂则一真一假。阑额中段形似隆起，两端刻作假月梁状，当然实际上它并不是弓形的。这种做法尚可见于其他少数元代建筑。

类似实例尚有河北曲阳县北岳庙的德宁殿（公元 1270 年）［元至元七年］（图 45）和山西赵城县广胜寺水神庙明应王殿，约公元 1320 年［元延祐七年，山西赵城县现已划入洪洞县］（图 46）。后者饰有壁画，作于公元 1324 年［元泰定元年］，其中有元代演剧场面。这种以世俗题材作为宗教建筑装饰的实例是很少见的。

山西［洪洞县］广胜寺的上、下二寺是两组使人感兴趣的建筑，它们全然不守常规（图 47）。在这些元末或明初的建筑中可见到巨大的昂，它们有时甚至被用以取代梁。这种结构也曾见于晋南的某些建筑，如临汾的孔庙，但在其他地区和其他时期的建筑中却不曾见，所以，也可能纯系一种地方特色。

浙江宣平县延福寺大殿（公元 1324—1327 年）［元泰定年间］，是长江下游和江南地区少见的一处元代建筑实例（图 48）。它那彻上露明造的梁架结构是复杂的大木作精品之一。它虽具有元代特征，但其柔和轻巧却与北方那些较为厚重的结构形成鲜明的对照。

地处西南的云南省也有少数元代建筑遗例。值得注意的是，这些边远省份的建筑虽然在整体比例上常常赶不上东部文化中心地区的演变，但在模仿当时建筑手法的某些细部方面却相当敏锐。这些建筑在整体比例上属于 12 或 13 世纪，在细部处理上却属于 14 世纪。

图 44

河北正定阳和楼
［已毁］

44

Yang-ho Lou, Cheng-
ting, Hopei, ca.1250
（Destroyed）

a. 全景
a. General view
b. 平面及断面图
b. Plan and cross
section

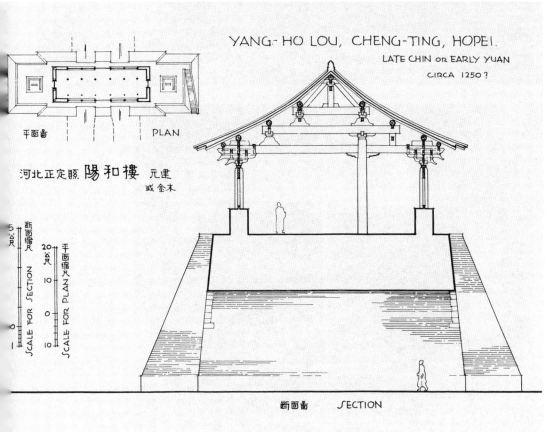

YANG-HO LOU, CHENG-TING, HOPEI.
LATE CHIN or EARLY YUAN
CIRCA 1250 ?

平面圖　　PLAN

河北正定縣 陽和樓 元建
或金末

斷面尺
5 公尺

SCALE FOR SECTION

平面縮尺
SCALE FOR PLAN

20
10
0
10

斷面圖　SECTION

图 45

河北曲阳北岳庙德宁殿

45

Te-ning Tien, Main Hall, Pei-yueh Miao, Ch'u-yang, Hopei,1270

a. 全景

a. General view

b. 下檐斗栱

b. Interior view of *tou-kung* supporting porch roof

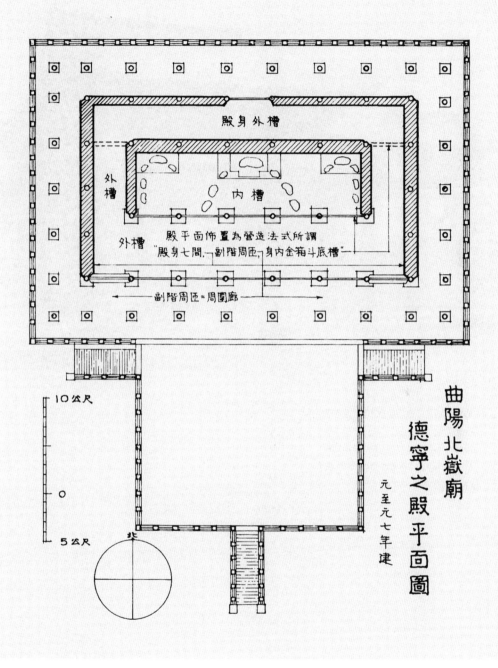

殿身外槽

外槽

内槽

外槽

殿平面佈置為營造法式所謂
"殿身七間，副階周匝，身内金箱斗底槽"

副階周匝＝周圍廊

10公尺

0

5公尺

北

曲陽 北嶽廟
德寧之殿平面圖
元至元七年建

c. 平面图

c. Plan

103

图 46
山西赵城县广胜寺明应王殿［水神庙］
　46
Ming-ying-wang Tien（Temple of the
Dragon King）, Kuang-sheng Ssu, Chao-
ch'eng, Shansi, ca.1320）

c. 下寺前殿梁架，
可见巨大的昂
c. Lower Temple
interior, showing
huge *ang*

d. 上寺前殿

d. Upper Temple hall

e. 上寺前殿梁架，可见巨大的昂

e. Upper Temple interior, showing huge *ang*

羁直时期

（约公元 1400—1912 年）

　　自 15 世纪初（明王朝）定都北京时起，主要在宫廷建筑中出现了一种与宋元时代迥然不同的风格。这种转变来得很突然，仿佛某种不可抗拒的力量突然改变了匠师们的头脑，使他们产生了一种全然不同于过去的比例观。甚至在明朝的开国皇帝洪武年间（公元1368—1398 年），建筑还保留着元代形制。这种醇和遗风的最后范例，可以举出山西大同的城楼（公元 1372 年）［洪武五年］和鼓楼（可能也建于同年）以及四川省峨眉山飞来寺的飞来殿（公元 1391年）［洪武二十四年］。

　　在这个新都的建筑中，斗栱在比例上的突然变化是一望而知的（图 32）。在宋代，斗栱一般是柱高的一半或三分之一，而到了明代，它们突然缩到了五分之一。在 12 世纪以前，补间铺作从不超过两朵，而现在却增至四或六朵，后来甚至是七八朵。这些补间铺作不仅不能再以巧妙的出跳分担出檐的荷载，而且连它们本身也成了阑额［清称额枋］的负担。过去，阑额的功能是联系多于负重，现在却不得不加大尺寸以承受这额外的负担。普拍枋［清称平板枋］与下面的阑额已不再呈 T 字形，而是与后者同宽，有时甚至还要略窄，因为它上面那缩小了的斗栱中的纤小栌斗［清称坐斗］并不需要垫一条过宽的板材。

　　在斗栱本身的做法上，也有其他一些重大变化。由于大得不合比例的梁直接落到了柱头铺作上，那种带长尾的昂已经无用武之地了。而在那些从外观效果上需要昂的地方便一律用上了假昂。但是在补间铺作的内面，昂尾却被广泛用作一种装饰性而非结构性的构件。昂尾上增加了许多华而不实的附加雕刻装饰。特别是所谓三福云，在宋代本是偶尔用于偷心华栱中的一根简单的纵向翼状构件，现在却发展为附在昂尾上的云朵。这种昂尾也不再是下端成喙状而斜置着的下昂的上端了。现在的昂嘴已是华栱的延伸部分，成为一种假昂。而昂尾则是一些平置构件如耍头或栱枋头（华栱上面的一根小枋）后面的延伸部分。当这种平置构件伸出一根往上翘的长尾时，它看起来有点像是曲棍球棍。它们已不再是支承檐檩的杠杆，反而成了一种累赘，要用外加的枋来支承。这时的斗栱，除了柱头铺作而外，已成了纯粹的装饰品。

在支承屋顶的梁架中，已完全不用斗栱。尺寸比过去加大了的梁现已直接安放在柱顶或瓜柱上，檩则直接由梁头来支撑，不再借助于栱，也不用叉手或托脚支撑；而脊檩的荷载完全由壮实的侏儒柱来承担。

柱的分配非常规则，以致建筑的平面变成了一个棋盘形。在约公元 1400 年以后，极少为了实用目的而抽减柱子以加大空间的做法。

永乐皇帝的陵墓（明长陵）

这种形式的建筑现存最早的实物是河北省昌平县明十三陵内明成祖［永乐］的长陵祾恩殿，建于 1403—1424 年间［永乐］。明代后继的诸帝后都葬在这一带，但长陵的规模最大，独据中央。

祾恩殿（图49）为陵寝的主要建筑，九间重檐，下有三层白石台基。它几乎完全仿照永乐帝在皇宫中听政的奉天殿（见下文）而建。其斗栱在比例上极小，但昂尾却特长。补间铺作有八攒之多，都是纯装饰性的。在这一时期的最初阶段，这么小的斗栱和这么多的补间铺作都是少见的。然而，这座建筑的整个效果还是极其动人的。

图 49
河北昌平明长陵祾恩殿

49
Sacrificial Hall, Emperor Yung-lo's Tomb, the Ming Tombs, Ch'ang-p'ing, Hopei, 1403-1424

c. 平面及断面图（右页下）
c. Plan and cross section

a. 全景

a. General view

b. 藻井及梁架细部

b. Interior detail of ceiling and structure

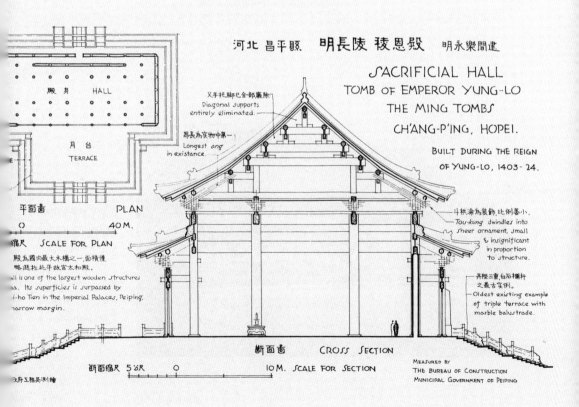

河北昌平縣　明長陵稜恩殿　明永樂間建

SACRIFICIAL HALL
TOMB OF EMPEROR YUNG-LO
THE MING TOMBS
CH'ANG-P'ING, HOPEI.

BUILT DURING THE REIGN
OF YUNG-LO, 1403-24.

殿身　HALL

月台
TERRACE

平面畫　PLAN

0　　　　　　40 M.

縮尺　SCALE FOR PLAN

殿為國內最大木構之一，面積僅
略遜於北平故宮太和殿。
...ll is one of the largest wooden structures
...a. Its superficies is surpassed by
...i-ho Tien in the Imperial Palaces, Peiping.
...narrow margin.

又手托脚巳全部廢除
Diagonal supports
entirely eliminated.

昂長為實物中第一
Longest ang
in existance.

斗栱淪為裝飾，比例甚小。
Tou-kung dwindles into
sheer ornament, small
& insignificant
in proportion
to structure.

丹陛三重白石欄杆
之最古實例。
Oldest existing example
of triple terrace with
marble balustrade.

斷面畫　CROSS SECTION

斷面縮尺 5 公尺　0　　　　10 M. SCALE FOR SECTION

MEASURED BY
THE BUREAU OF CONSTRUCTION
MUNICIPAL GOVERNMENT OF PEIPING

...府工務局測繪

北京故宫中的明代建筑

北京的明代皇宫是在元朝京城，即马可·波罗曾经访问过的元大都的劫后废墟上重建起来的。尽管已有五个半世纪之久，但其基本格局至今变化不大。虽然皇帝临朝的主要大殿皇极殿［明初称为奉天殿，清代改称为太和殿］在明朝败亡时曾被毁，但宫内仍有不少明代建筑。其中，建于公元1421年［明永乐十九年］的社稷坛享殿——在今中山公园内——是最早的一处（图50）。在这座建筑上，斗栱仍约为柱高的七分之二，当心间只有补间铺作六攒，其余各间则只有四攒。宫内另一组值得注意的建筑是重建于公元1545年［明嘉靖二十四年］的太庙，即祭祀皇族祖先的庙宇（图51）。

故宫三大殿中的最后一座，原名建极殿［后改称保和殿］，是公元1615年［明万历四十三年］在一次火灾后重建的（图52）。1644年明朝覆灭时故宫遭焚，1679年［清康熙十八年］故宫又一次失火，三大殿的前面两座均被毁，独建极殿幸免于难。此殿与清《工程做法则例》（1734年）［雍正十二年］时期的其他建筑，无论

图51
北京皇城内太庙（右页）
51
T'ai Miao, Imperial Ancestral Temple, Imperial Palace, Forbideen City, Peking. Rebuilt 1545

图50
北京皇城内社稷坛
［今中山公园］享殿［今中山堂］
50
Hsiang Tien, Sacrificial Hall, She-chi T'an, Forbidden City, Peking,1421

图 52

北京故宫保和殿
［原图注"后来曾
重建"，有误］

52
Pao-ho Tien（for-
merly Chien-chi
Tien），Imperial
Palace, Peking,
1615. Later rebuilt

在整体比例上还是在细节上都大体相同，以致若非在藻井以上发现了每一构件上都有以墨笔标明的"建极殿"某处用料字样，人们是很难确认它为清代以前遗构的。

山东曲阜孔庙奎文阁（公元 1504 年）[明弘治十七年]，高二层，是明代官式做法的一个引人注意的实例（图 53）。但是，看来这种羁直的风格影响所及并没有超出北京多远。确切地说，并未超出按宫廷命令和官式制度兴建的那些建筑的范围。在清帝国的其他地区，匠师们要比宫廷建筑师自由得多。全国到处都可以见到那种多少仍承袭着旧传统的建筑物，如四川梓潼县文昌宫内的天尊殿[建于明中叶]和四川蓬溪县鹫峰寺的建筑群（始建于公元 1443 年）[明正统八年]都是其中杰出的实例。

图 53

山东曲阜孔庙奎文阁

53

Library, Temple of Confucius, Ch'u-fu. Shantung,1504

a. 全景

a. General view

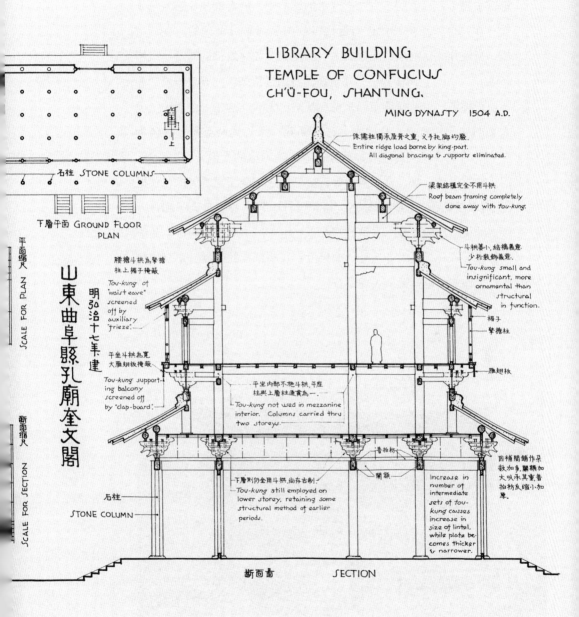

LIBRARY BUILDING
TEMPLE OF CONFUCIUS
CH'Ü-FOU, SHANTUNG.

MING DYNASTY   1504 A.D.

石柱 STONE COLUMNS

下層平面 GROUND FLOOR
PLAN

平面縮尺 SCALE FOR PLAN

斷面縮尺 SCALE FOR SECTION

山東曲阜縣孔廟奎文閣

明弘治十七年建

侏儒柱獨承脊之重，乂手扶脚均廢.
Entire ridge load borne by king-part.
All diagonal bracings & supports eliminated.

梁架結構完全不用斗栱
Roof beam framing completely
done away with tou-kung.

斗栱菱小，結構意意.
少作裝飾意意.
Tou-kung small and
insignificant, more
ornamental than
structural
in function.

楊子
擎檐柱

雕翅栿

腰檐斗栱為擎檐
柱上楊子掩蔽
Tou-kung of
"waist eave"
screened
off by
auxiliary
"frieze".

平坐斗栱為覆
大雁翅板掩蔽
Tou-kung support-
ing balcony
screened off
by "clap-board".

平坐內部不施斗栱，平坐
柱與上層柱通貫為一.
Tou-kung not used in mezzanine
interior. Columns carried thru
two storeys.

普拍枋

闌額

石柱
STONE COLUMN

下層則仍全用斗栱，尚存古制.
Tou-kung still employed on
lower storey, retaining some
structural method of earlier
periods.

因補間鋪作朵
數加多，闌額加
大以承其重普
拍枋反縮小，加
厚.
Increase in
number of
intermediate
sets of tou-
kung causes
increase in
size of lintel,
while plate be-
comes thicker
& narrower.

斷面圖   SECTION

北京的清代建筑

　　清代（公元 1644—1912 年）的建筑只是明代传统的延续。在公元 1734 年［雍正十二年］《工程做法则例》刊行之后，一切创新都被窒息了。在清朝二百六十八年的统治中，所有的皇家建筑都千篇一律，这一点是其他任何近代极权国家都难以做到的。在紫禁城、皇陵和北京附近的无数庙宇中的绝大多数建筑都同属这一风格（图 54—图 57）。它们作为单个建筑物，特别是从结构的观点来看，并不值得称道；但从总体布局来说，却举世无双。这是一个规模上硕大无朋的宏伟布局。从南到北，贯穿着一条长约两英里（三公里左右）的中轴线，两边对称地分布着绵延不尽的大道、庭院、桥、门、柱廊、台、亭、宫、殿等等，全都按照完全相同的、严格根据《工程做法则例》的风格建造，这种设计思想本身就是天子和强大帝国的最适当的表现。在这种情况下，由于严格的规则而产生的统一性成了一种长处而非短处。如果没有这些刻板的限制，皇宫如此庄严宏伟之象也就无从表现了。

图 54
北京故宫西华门
54
Hsi-hua Men, a gate of the Forbidden City, Peking, Ch'ing dynasty

图 55
北京故宫角楼
55
Corner tower of
the Forbidden City,
Peking, Ch'ing
dynasty
a. 全景
a. General view

b. 外檐斗栱
b. Exterior *tou-kung*

117

图 56

北京故宫文渊阁，
建于 1776 年［清
乾隆四十一年］

56

Wen-yuan Ke,

Imperial Library,

Forbidden City,

Peking, 1776

a. 正面细部

a. Detail of facade

b. 平面及断面图

b. Plan and cross

section

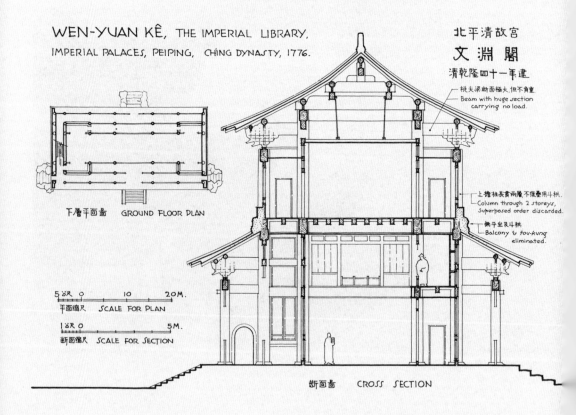

WEN-YUAN KÊ, THE IMPERIAL LIBRARY,
IMPERIAL PALACES, PEIPING, CHING DYNASTY, 1776.

北平清故宫
文淵閣
清乾隆四十一年建

桃夫梁断面極大,但不負重
Beam with huge section
carrying no load.

上檐柱貫書兩層不復疊用斗栱
Column through 2 storeys,
Superposed order discarded.

無平坐及斗栱
Balcony & tou-kung
eliminated.

下層平面畫 GROUND FLOOR PLAN

5 尺 0   10   20M.
平面縮尺 SCALE FOR PLAN

1 尺 0   5M.
断面縮尺 SCALE FOR SECTION

断面畫 CROSS SECTION

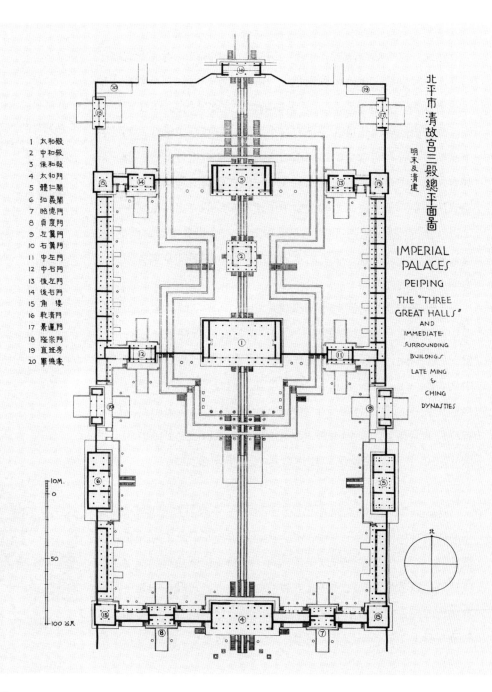

北平市清故宫三殿總平面圖

明末及清建

1 太和殿
2 中和殿
3 保和殿
4 太和門
5 體仁閣
6 弘義閣
7 昭德門
8 貞度門
9 左翼門
10 右翼門
11 中左門
12 中右門
13 後左門
14 後右門
15 角　樓
16 乾清門
17 景運門
18 隆宗門
19 直班房
20 軍機處

IMPERIAL
PALACES
PEIPING
THE "THREE
GREAT HALLS"
AND
IMMEDIATE·
SURROUNDING
BUILDINGS
LATE MING
&
CHING
DYNASTIES

北

图 57

北京明清故宫三殿总平面图

57

Imperial Palaces, Forbidden City, Peking,
Ming and Ch'ing dynasties. Site plan

然而，这个布局却有一个重大缺陷，看来其设计者完全忽略了那些次要横轴线，也许可以说是无力解决这个问题。甚至就在主轴线上各殿两侧布置建筑时，其纵横轴线之间的关系也往往不甚协调。故宫内的各建筑群几乎都有这个问题，特别是中轴线两侧的各个庭院，尽管在它本身的四面围墙之内是平衡得很好的——每一群都有一条与主轴线平行的南北轴线——但在横向上却与主轴线没有明确的关系。虽然强调主轴线是中国建筑布局的突出特征之一，这可以从全国所有的庙宇和住宅平面中看出。但几乎难以置信的是，设计者怎能如此重视一个方向上轴线的对称，同时却全然无视另一方向上轴线的处理。

太和殿是故宫中主要的听政殿，也是整座皇城的中心，是各殿中最宏伟的一座单幢建筑（图58）。它共有六行柱子，每行十二根，广十一间，进深五间，庑殿重檐顶，是中国现存古代单幢建筑中最大的一座。殿内七十二根柱子排列单调而规整，虽无巧思，却也颇为壮观。大殿建在一个不高的白石阶基上，下面则是三层带有栏杆的台阶，上面饰有极其精美的雕刻。殿为公元1679年［康熙十八年］火灾后重建，其建造年代不早于公元1697年［康熙三十六年］。

这座巨型大殿的斗栱在比例上极小——不及柱高的六分之一。当心间的补间铺作竟达八攒之多。从远处望去几乎见不到斗栱。大殿的墙、柱、门、窗都施以朱漆，而斗栱和额枋则是青绿描金。整座建筑覆以黄色琉璃瓦，在北方碧空的衬托下，它们在阳光中闪耀着金色的光芒。白石阶仿佛由于多彩的雕饰而激荡着，那雄伟的大殿矗立其上，如一幅恢宏、庄严、绚丽的神奇画面，光辉夺目而使人难忘。

这里顺便提一句，太和殿还装备着一套最奇特的"防火系统"。在当心间屋顶下黑暗的天花上，供着一个牌位，上面刻有佛、道两教的火神、风神和雷神的姓名和符咒，前面有香炉一尊、蜡烛一对，还有道教中象征长生不老的灵芝一对。看起来这个预防系统似

图58
北京故宫太和殿
58
T'ai-ho Tien, Hall
of Supreme Harmo-
ny, Forbidden City,
Peking, 1697
a. 全景
a. General view
b. 白石台基
b. Marble terraces
c. 藻井
c. Ceiling

图像中国建筑史

乎还很灵验呢!

清代的另一座著名建筑是天坛的祈年殿。殿为圆形，重檐三层，攒尖顶（图59）。蓝琉璃瓦象征着天的颜色。这座美丽的建筑也是建在三层有栏杆的白石台基之上的。现在这座大殿是公元1890年［清光绪十六年］重建的，原先的一座于此前一年被焚毁。

北京的护国寺是明清时代一座典型的佛教寺庙。它始建于元代，但在清代曾彻底重建。寺内建筑的布局可使各进庭院由建筑的两侧互通。寺庙中特有的钟、鼓二楼被置于前院的两侧。全寺最后的一座元代建筑现已完全坍毁，却是木骨土墼墙中一个令人感兴趣的实例。

明、清两代还建造了许多清真寺。但除内部装修的细部之外，它们与其他建筑并无本质上的区别。

图59
北京天坛祈年殿
　　59
Ch'i-nien Tien,
Temple of Heaven,
Peking, 1890

山东曲阜孔庙

山东曲阜县孔庙大成殿（图60）的建成年代（公元1730年）[雍正八年] 与《工程做法则例》的颁行几乎同时，然而却与书中规定颇有出入。它的雕龙石柱虽然很壮观，但整个比例却显得有些生硬，为清代其他建筑中所未见，无论那些建筑是否由皇帝敕令所建。

曲阜孔庙自汉代以降就属国家管理，是中国唯一一组两千余年未间断的历史的建筑物。现在孔庙是一个巨大的建筑群，占据了县城的整个中心区，其布局是宋代时定下的。围墙之内包括了不同时期的众多建筑，可谓五光十色。其中最早的是建于金代，即公元1195年 [明昌六年，原文公元1196年有误] 的碑亭，最晚的则建于1933年。其间的元、明、清三代，在这里都有建筑遗存。

图 60
山东曲阜孔庙
60
Temple of Con-
fucius, Ch'ü, Shan-
tung,1730
a. 大成殿正面
a. Ta-ch'eng Tien,
Main Hall, facade

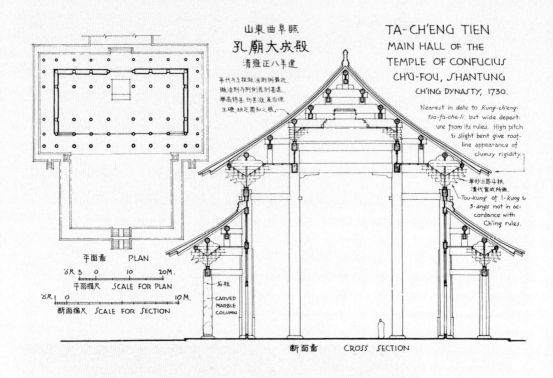

山東曲阜縣
孔廟大成殿
清雍正八年建

筆代與三探做法最近
做法則與則例差別甚違
舉高特甚折甚激屋面線
生硬缺乏圓和之感

TA-CH'ENG TIEN
MAIN HALL OF THE
TEMPLE OF CONFUCIUS
CHÜ-FOU, SHANTUNG
CH'ING DYNASTY, 1730.

Nearest in date to Kung-ch'eng-
tso-fa-che-li, but wide departure from its rules. High pitch
& slight bent give roof-
line appearance of
clumsy rigidity.

單抄三昂斗拱
清代官式所無
Tou-kung of 1-kung &
3-angs not in ac-
cordance with
Ch'ing rules.

平面圖 PLAN

ठR 5   0   10   20M.
平面縮尺 SCALE FOR PLAN

ठR 1 0   10 M.
斷面縮尺 SCALE FOR SECTION

石柱
CARVED
MARBLE
COLUMN

斷面圖   CROSS SECTION

STELE PAVILION, TEMPLE OF CONFUCIUS
CHÜ-FOU, SHAN-TUNG, CHIN DYNASTY
1196 A.D.

此線以上部分清乾隆間改修
Portion above this level
rebuilt in 18th century.

石柱
Stone columns

正心枋正心桁及桁
搅為清官式做法
18th century alteration,
rest of tou-kung
original.

平面圖 PLAN

山東曲阜縣 孔廟碑亭
金明昌六年建?

為聖廟現存最古建築後世重修頗
有更改尤以上層屋頂梁架為
甚拱類斗拱則大
部仍保持原狀

Oldest wooden
structure in the Sage's
Temple. Top portion sup-
porting ridge & roof much
altered by later repairs.
Columns, lintels & tou-kungs
are mostly original.

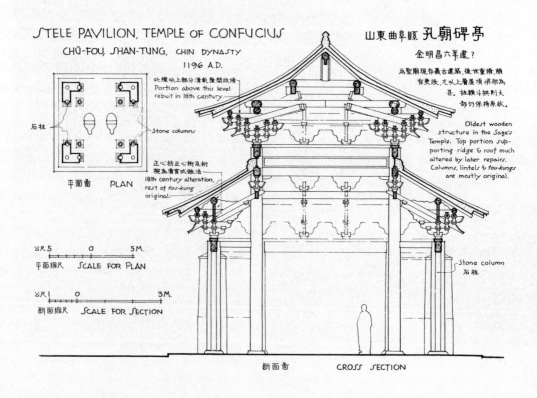

ठR 5   0   5M.
平面縮尺 SCALE FOR PLAN

ठR 1   0   3M.
斷面縮尺 SCALE FOR SECTION

Stone column
石柱

斷面圖   CROSS SECTION

b. 大成殿平面及断面图（左页上）

b. Main Hall, plan and cross section

c. 全庙平面图

c. Plan of entire compound

d. 碑亭平面及断面图（左页下）

d. Stela pavilion, plan and cross section, 1196

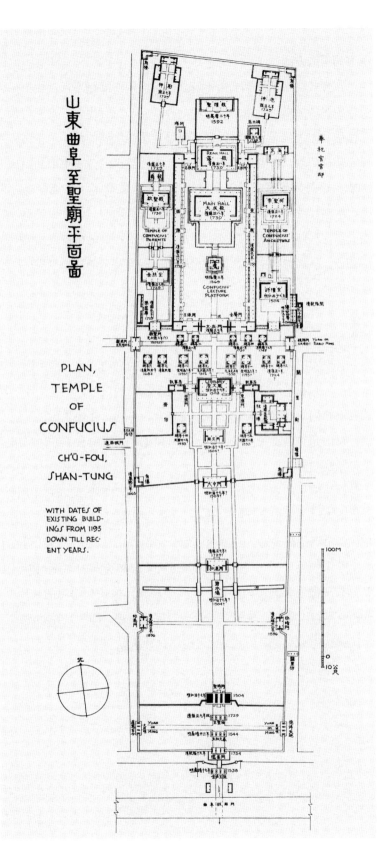

山東曲阜至聖廟平面圖

PLAN,
TEMPLE
OF
CONFUCIUS
CH'Ü-FOU,
SHAN-TUNG

WITH DATES OF
EXISTING BUILD-
INGS FROM 1195
DOWN TILL REC-
ENT YEARS.

南方的构造方法

在官式建筑则例影响所不及之处，即使离北京不远，由于采取了较为灵巧的做法，也使建筑物的外观看来更有生气。这种现象在江南诸省尤为显著。这种差异不仅是较暖的气候使然，也是南方人匠心巧技所致。在温暖的南方地区，无须厚重的砖、土墙和屋顶来防寒。板条抹灰墙，椽上直接铺瓦，连望板都不用的建筑随处可见。木料尺寸一般较小，屋顶四角常常高高翘起，颇具愉悦感。然而，当这种倾向发展得过分时，常会导致不正确的构造方法和繁缛的装饰，从而损害了一栋优秀建筑物所不可缺少的两大品质——适度和纯朴。

住宅建筑

云南省的民居看来倒很巧妙地把南方的灵巧和北方的严谨集于一身了。它在平面布局上有某种未见于他处的灵活性，把大小、功能各不相同的许多单元运用自如地结合在一起，并使其屋顶纵横交错。窗的布局也富于浪漫色彩，上有窗檐，下有窗台，在体形组合上极具画意。

就地取材的农村建筑，如人们在浙江武夷山区的农村中所见，在江南地区是有代表性的（图61）。但在北方黄土高原地区，依土崖挖成的窑洞仍很普遍。在滇西滇缅公路沿线山区，有一种特别的木屋［井干构民居］，其构造与斯堪的纳维亚和美洲的木屋相似，但又具有某种地道的中国特色，尤其表现在其屋顶和门廊的处理上。这使人不得不承认，建筑总是渗透着民族精神，即使是在如此偏远地区偶然建造的简陋小屋，也表现出这种情况（图62）。

图 61
浙江武夷山区民居
61
Mountain homes,
Wu-i, Chekiang

图 62
云南镇南县马鞍山
井干构民居图
62
Log cabin, western
Yunnan, plans and
elevations

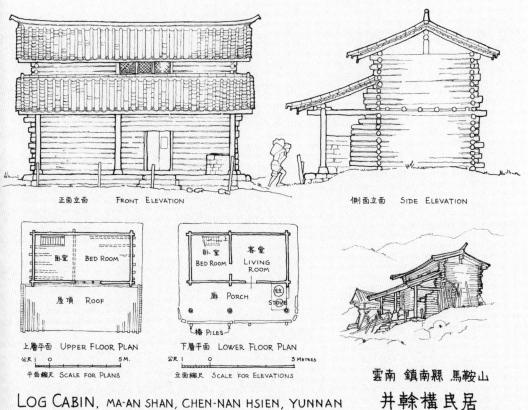

正面立面 FRONT ELEVATION　　　　　　侧面立面 SIDE ELEVATION

卧室 BED ROOM

屋顶 ROOF

上层平面 UPPER FLOOR PLAN

公尺 1　0　　　5 M.
平面缩尺 SCALE FOR PLANS

卧室 BED ROOM

客堂 LIVING ROOM

廊 PORCH　灶 STOVE

椿 PILES

下层平面 LOWER FLOOR PLAN

公尺 1　0　　3 METRES
立面缩尺 SCALE FOR ELEVATIONS

LOG CABIN, MA-AN SHAN, CHEN-NAN HSIEN, YUNNAN

RED BY LIU T.T.

雲南 鎮南縣 馬鞍山
井幹構民居

劉敦楨 測繪

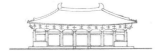

佛塔

在表现并点缀中国风景的重要建筑中，塔的形象之突出是莫与伦比的，从开始出现直至今日，中国塔基本上是如上文曾引述的"下为重楼，上累金盘"，也就是这两大部分——中国的"重楼"与印度的窣堵坡（"金盘"）的巧妙组合。依其组合方式，中国塔可分为四大类：单层塔、多层塔、密檐塔和窣堵坡。不论其规模、形制如何，塔都是安葬佛骨或僧人之所。

根据前引的那类文献记载和云冈、响堂山、龙门石窟（图17—图19）中所见佐证，以及日本现存的实物，可以看出早期的塔都是一种中国本土式的多层阁楼，木构方形，冠以窣堵坡，称为刹。但匠师们不久就发现用砖石来建造这类纪念性建筑的优越性，于是便出现了砖石塔，并终于取代了其木构原型。除应县木塔（图31）这唯一遗例之外，中国现存的塔全部为砖石结构。

砖石塔的演变（图63）大致可分为三个时期：古拙时期，即方形塔时期（约公元500—900年）；繁丽时期，即八角形塔时期（约公元1000—1300年）；以及杂变时期（约公元1280—1912年）。与我们对木构建筑的分期相似，这种分期在风格和时代特征上必然会有较长时间的交叉或偏离。

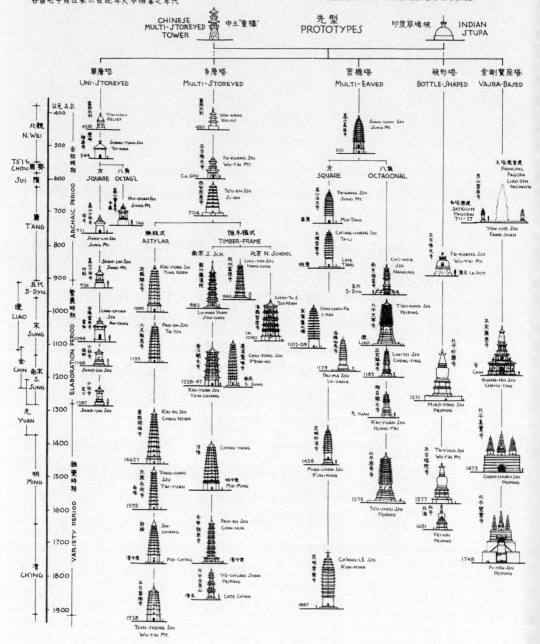

歷代佛塔型類演變圖　EVOLUTION OF TYPES OF THE BUDDHIST PAGODA

各圖非用同一縮尺　NOT DRAWN TO SAME SCALE
附畫人像以示塔約略大小　HUMAN FIGURE INDICATES APPROXIMATE SCALE
各圖地平經位表示在紀年尺中相當之年代　POSITION OF GROUD-LINE INDICATES DATE IN RELATION TO CHRONOLOGICAL SCALE.

132

古拙时期

（约公元 500—900 年）

　　这一时期大体始自 6 世纪初，直到 9 世纪末。其间历经北魏、北齐、隋、唐诸代。其显著特征，除少数例外，均为方形、空心单筒，即塔成筒状，内部不再用砖构分层分间（但可能有木制楼板、楼梯），如同一个封了顶的近代工厂的大烟囱。上述四种形式的塔中，前三种已在此期间内出现，并有很多实例。唯有塔的先型印度式窣堵坡，料想在此早期会有实物，却不曾见；尽管完全有理由相信中国人此时已经知道了这一形式。这一点令人不解。

单层塔

　　除一例之外，单层塔都是僧人的墓塔。它们规模不大，看起来更像神龛而不像通常所说的塔。在云冈石窟浮雕中，这种塔的形象甚多。其特征是一方形小屋，一面有拱门，上面是一或两层屋檐，再上覆以刹。

　　山东济南附近神通寺四门塔（公元 611 年）［隋大业七年］是中国最早的石塔，也是现存单层塔中最早和最重要的一座（图 64a）。但它绝非这类塔的典型，因为它既不是墓塔，又不是空筒结构，而是一座方形单层亭状石砌建筑。中央为一方墩，四周贯通，四方各有一孔券门。在这一时期，塔的内部做如此处理的仅此孤例。但在10 世纪以后，这却成了塔的普遍形式。

　　山东长清县灵崖寺慧崇禅师塔（约公元 627—649 年）［唐贞观年间］（图 64b）和建于公元 771 年［唐大历六年］的河南登封县嵩

图 63
历代佛塔型类演变图

63
Evolution of types of the Buddhist Pagoda

图 64   a. 山东济南附近神

单层塔   通寺四门塔

   64   a. Ssu-men T'a,

One-storied Pagodas Shen-t'ung Ssu,

            near Tsinan,

            Shantung

山少林寺同光禅师塔（图64c）是这类单层墓塔的典型实例。

　　河南登封县嵩山会善寺净藏禅师塔是在建筑方面具有极重要意义的一个独特的典型（图64d，图64e），塔建于禅师圆寂（公元746年）[唐天宝五年]后不久，为一栋较小的单层八角形亭式砖砌建筑，下面有一个很高的须弥座。塔身外面在转角处砌出倚柱，并有斗栱、假窗及其他构件。斗栱形制与云冈及天龙山石窟中所见相近，但在每一栌斗处伸出一根与栱相交的要头。八面阑额上各有一朵人字形补间铺作。整座塔形为当时的典型形制；但八角形平面和须弥座却是第一次出现，而自10世纪中叶以后，这两者已成为塔的两项显著特征。然而，在8世纪中期的建筑上采用这些做法，却是塔形演进中将发生重大变化的先兆。

b. 山东长清灵崖寺慧崇禅师塔

b. Tomb Pagoda of Hui-ch'ung, Ling-yen Ssu, Ch'ang-ch'ing, Shantung, ca.627–649

c. 河南登封嵩山少林寺同光禅师塔

c. Tomb Pagoda of T'ung-kuang, Shao-lin Ssu, Teng-feng, Honan,771

d. 河南登封嵩山会
善寺净藏禅师塔

d. Tomb Pagoda of
Ching-tsang, Hui-
shan Ssu, Teng-
feng, Honan, 746

e. 会善寺净藏禅师
塔平面图

e. Tomb Pagoda of
Ching-tsang, plan

河南登封照會善寺净藏禅師塔
平面圖

北

PLAN, CHING-TSANG CH'AN-SHIH T'A,
HUI-SHAN SSU, TENG-FENG, HONAN.
MEASURED BY LIU T.T.
劉敦楨測繪

f. 少林寺行钧禅
师塔，建于 926 年
［后唐天成元年］
f. Tomb Pagoda of
Hsing-chun, Shao-
lin Ssu, 926

多层塔

多层塔是若干单层塔的叠加。这是云冈石窟浮雕及圆雕中最常见的一种塔。各层的高度和宽度往上依次略减。现存砖塔外表常饰以稍稍隐起的倚柱，柱端有简单的斗栱，是对当时木构建筑的模仿。陕西西安慈恩寺大雁塔（图65a，图65b）是这种形式的塔中最著名的一座。其原构是7世纪中叶玄奘法师所献，不久即毁，今塔建于公元701—704年［唐武后长安间］。这是一座典型的空筒状塔，内部楼板及楼梯都是木构。塔面以十分细致的浮雕手法砌出非常瘦长的扁柱，与豪劲粗大的塔身恰成鲜明对比。各柱上均仅有一斗，没有补间铺作。各层四面都开有券门，底层西门门楣上有一块珍贵的石刻，描绘了一座唐代的木构大殿（图22）。

在这种形式的塔中，还有两个应当提到的例子，即西安市附近的香积寺塔和玄奘塔。前者建于公元681年［唐永隆二年］，在总体布局和外墙处理上与大雁塔相似，但其补间也用单斗，墙上还有假窗（图65c）。兴教寺内的玄奘塔建于公元669年［唐总章二年］，是个只有五层的小塔，可能是这位大师的墓塔。[1] 塔的底层外墙素平无柱，其上四层则隐起倚柱。在斗栱处理上，它与净藏塔相近，栌斗中有耍头伸出，但耍头外端呈直面。这座塔比净藏塔早约七十年，是采用这种做法的最早实例（图65d）。

山西五台山佛光寺中称为祖师塔的那座多层塔，是一个极特殊的实例（图65e）。它位于该寺唐代大殿南面，与殿相距不过几步。六角形，两层，下层上部砌斗一圈，斗上为莲瓣檐，再上为叠涩檐；上层之下为平坐，平坐为须弥座形，用版柱将仰莲座与下涩之间的束腰隔成若干小格。三檐均用莲瓣。这种须弥座做法虽属细节，却是6世纪中后期最典型的形式，甚至可用以作为判断年代的一种标志。如前所述，在多层木构建筑中，平坐具有重要作用。因此，此处出现的平坐颇值得重视（图65f）。

祖师塔的上层较富建筑意味。转角处都砌出倚柱，柱的两端及中间都以束莲装饰，显然系受印度影响。正面墙上还砌有假门，门

〔1〕
原文中所注这两塔的建造年代应互易，译文已根据作者所著《中国建筑史》改正。——孙增蕃校注

图 65
多层塔
65
Multi-storied Pago-
das
a. 陕西西安慈恩寺
大雁塔
a. Ta-yen T'a (Wild
Goose Pagoda),
Tzu-en Ssu, Sian,
Shensi, 701-704
b. 大雁塔平面图
b. Ta-yen T'a, plan

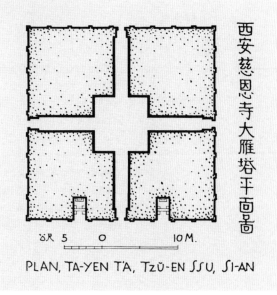

西安慈恩寺大雁塔平面图

ŏR 5　　0　　10 M.

PLAN, TA-YEN TA, TZŬ-EN SSU, SI-AN

c. 陕西西安香积寺塔
c. Hsiang-chi Ssu T'a,
Sian, Shensi, 669

d. 陕西西安兴
教寺玄奘塔　　d. Hsuan-tsang T'a, Hsing-chiao
　　　　　　　Ssu, Sian, Shensi, 681

e. 山西五台山佛光寺
祖师塔，约建于公元
600 年
e. Tsu-shih T'a, Fo-kuang
Ssu, Wu-tai, Shansi,
ca.600
f. 佛光寺祖师塔平面及
立面图（右页）
f. Tsu-shih T'a, plan and
elevation

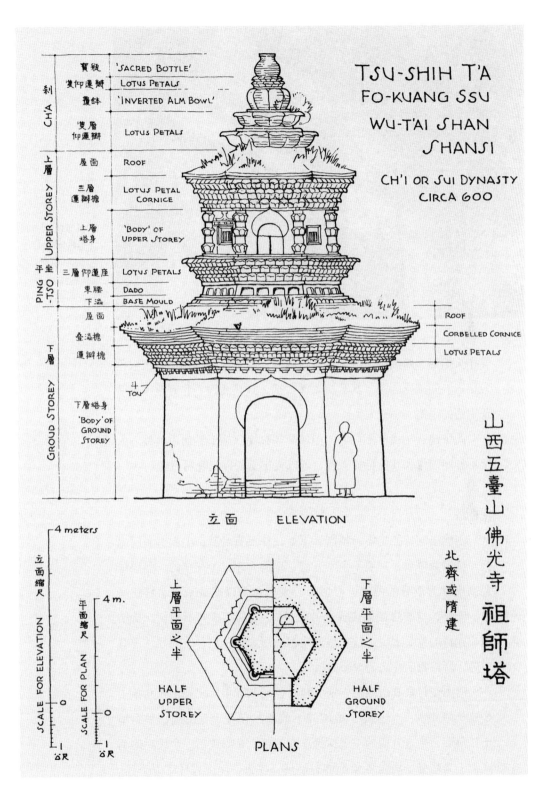

寶瓶 'SACRED BOTTLE'
覆仰蓮瓣 LOTUS PETALS
覆鉢 'INVERTED ALM BOWL'

剎 CHA

覆層仰蓮瓣 LOTUS PETALS

上層 UPPER STOREY

屋面 ROOF
三層蓮瓣檐 LOTUS PETAL CORNICE
上層塔身 'BODY' OF UPPER STOREY

平坐 PING-TSO
三層仰蓮座 LOTUS PETALS
束腰 DADO
下澀 BASE MOULD

下層 GROUND STOREY
屋面 ROOF
疊澀檐 CORBELLED CORNICE
蓮瓣檐 LOTUS PETALS

斗 TOU

下層塔身 'BODY' OF GROUND STOREY

TSU-SHIH T'A
FO-KUANG SSU
WU-T'AI SHAN
SHANSI

CH'I OR SUI DYNASTY
CIRCA 600

山西五臺山 佛光寺 祖師塔

北齊或隋建

立面 ELEVATION

立面縮尺 SCALE FOR ELEVATION

4 meters

平面縮尺 SCALE FOR PLAN

4 m.

0

0

1 ÖR

1 ÖR

上層平面之半
HALF UPPER STOREY

下層平面之半
HALF GROUND STOREY

PLANS

g. 祖师塔二层窗的
上方所绘斗栱
g. Tsu-shih T'a,
painted *tou kung*
over secong-story
window

券作火焰形，两侧墙上则砌出假直棂窗，窗上白墙仍残留着土朱色
人字形补间铺作画迹，其笔法雄浑古拙，颇具北魏、北齐造像衣褶
及书法的风格（图 65g）。

此塔建造年代已不可考。但从其须弥座、焰形券面束莲柱、人
字形补间铺作和其他特征来看，可以肯定是 6 世纪晚期遗构。

密檐塔

密檐塔的特征是塔身很高，下面往往没有台基，上面有多层出
檐。檐多为单数，一般不少于五层，也鲜有超过十三层的。各层檐
总高度常为塔身的两倍。习惯上，人们总是以檐数来表示塔的层
数，于是，这类塔便被说成是"几层塔"，其实这并不确切。从结
构或建筑的意义上说，这类塔的出檐一层紧挨一层，中间几乎没有
空隙，所以我们称之为"密檐塔"。

河南登封县嵩山嵩岳寺塔，建于公元 520 年［北魏正光元年］，
是这类塔中的一个杰作，虽然并不典型。它有十五层檐，平面呈
十二角形，是这方面的一个孤例。它的整个布局包括一个极高的基
座，上为塔身，塔身各角都饰有倚柱，柱头仿印度式样饰以垂莲。

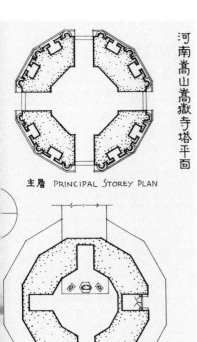

河南嵩山嵩嶽寺塔平面

主層 PRINCIPAL STOREY PLAN

基層 GROUND STOREY PLAN

AGODA OF SUNG-YÜEH SSU
G MOUNTAINS · TENG-FENG · HONAN

測繪　　　MEASURED BY LIU, T.-T.

图 66　　　　塔平面图

密檐塔　　　a. Plan of Sung-yuen

66　　　　　Ssu T'a

Multi-eaved Pagodas　b. 河南登封嵩岳

a. 河南登封嵩岳寺　寺塔

b. Sung-yuen Ssu

T'a, Teng-feng, Sung

Shan, Honan, 520

塔壁四面开有券门，其余八面各砌成单层方塔形的壁龛，凸出于塔壁外，龛座上刻有狮子。各层出檐依一条和缓的抛物线向内收分，使塔的轮廓显得秀美异常。全塔是一个砖砌空筒，内部平面为八角形（图66a，图66b）。塔内原有的木制楼板和楼梯已毁，这使它从内部仰望竟像一个电梯井。

这一时期典型的密檐塔都呈正方形，下无台基，塔身外墙多为素面。这种塔最好的实例是建于8世纪早期的河南登封县嵩山永泰寺塔和法王寺塔（图66c，图66d）。

陕西西安荐福寺小雁塔建于公元707—709年［唐景龙年间］[1]，也属于这一类型，但在其檐与檐之间狭窄的墙面上有窗（图66e，图66f）。然而其出檐，尤其是上层出檐，与其内部的分层并不一致，因此不能表示其内部的层次。其主层与以上各"层"在高度上相差悬殊，这与多层塔有规则的分层迥然不同。所以必须正确地将它列入密檐塔一类。建于南诏时的云南大理县佛图寺塔（公元820年）和千寻塔（约公元850年）也同属这一类型（图66g—图66j）。

有不少唐代石塔也属于密檐型。它们一般较小，高度很少超过25—30英尺［8—9米］。其出檐用薄石板，刻成阶梯状以模仿砖塔叠涩。主层的门多为拱门，券面作焰形，两侧并有金刚侍立。其典型遗例为河北［今北京市］房山县云居寺主塔旁的四座小石塔（公元711—727年）［原文年代有误，已改正］。它们位于一座大塔台基的四角，成拱辰之势（图66k，图66l）。这种五塔共一基座的形式后来在明清时期十分普遍。

〔1〕
原文所注系建寺年代而非建塔年代，译文按作者所著《中国建筑史》改正。——孙增蕃校注

窣堵坡

约在10世纪末，出现了一种比前述任何类型都更具印度色彩的塔。塔身近半球形的印度窣堵坡式墓塔，在敦煌石窟壁画中随处可见。这一时期中的实例虽在新疆地区有不少，但在中原一带却罕见，唯山西五台山佛光寺内有一例。然而，后来窣堵坡终究还是在中国站住了脚。此点在下文中论及杂变时期时再谈。

c. 河南登封永泰
寺塔

c. Yung-t'ai Ssu
T'a, Teng-feng,
Honan, eight
century

d. 河南登封法王
寺塔

d. Fa-wang Ssu T'a,
Teng-feng, Honan,
eight century

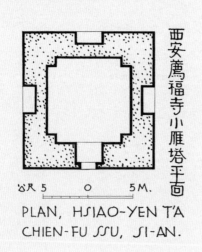

西安薦福寺小雁塔平面

PLAN, HSIAO-YEN T'A
CHIEN-FU SSU, SI-AN.

雲南大理縣佛圖寺塔平面

PLAN OF PAGODA
FO-T'U SSU
TA-LI · YUNNAN

e. 西安荐福寺小
雁塔
e. Hsiao-yen T'a,
Chien-fu, Sian,
Shensi, 707–709

f. 荐福寺小雁塔平
面图
f. Hsiao-yen T'a,
plan

g. 云南大理佛图
寺塔
g. Fo-t'u Ssu T'a,
Tali, Yunnan, 820(?)

h. 佛图寺塔平面图
h. Fo-t'u Ssu T'a,
plan

i. 云南大理千寻塔

i. Ch'ien-hsun T'a,

Tali, Yunnan, ca.850

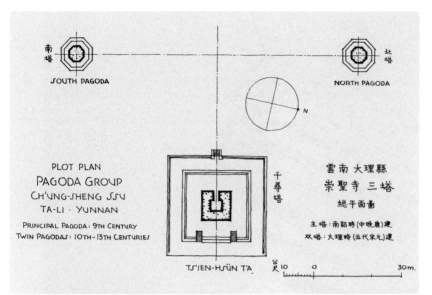

j. 云南大理崇圣寺
三塔总平面图
j. Ch'ien-hsun T'a
and twin pagodas,
Tali, site plan

k. 河北房山云居寺石塔，图中可见四座唐
代小石塔中之两座及主塔基座，主塔系辽
代所建

k. Stone Pagodas of Yun-chu Ssu, Fang Shan,
Hopei, showing two of four small T'ang
pagodas（711-727）and base of large
central pagoda, a Liao substitution

l. 房山云居寺唐代小石塔细部

l. Detail of a small T'ang pagoda, Yun-chu
Ssu, Fang Shan

繁丽时期

（约公元 1000—1300 年）

繁丽时期大体开始于 10 世纪末，结束于 13 世纪末，即五代、两宋以及辽金时期。这个时期塔的特征是：平面呈八角形，并开始用砖石在塔内砌出横向和竖向的间隔，形成回廊和固定的楼梯。这种间隔与过去的筒形结构相比，使塔的内观大异其趣。但这种构思也并非新创，因为早在 6 世纪中期，它就曾一度出现于神通寺的四门塔中（图 64a）。

现存最早的八角形塔是公元 746 年 [唐天宝五载] 所建净藏禅师塔（图 64d，图 64e），这也是第一次真正表达了英文中 Pagoda 一词按其读音来讲的确实含义。人们一直弄不清这个怪词的词源从何而来。看来最合理的解释是：那无非是按中国南方发音读出的汉字"八角塔"的音译而已。在本书 [英文本] 中，有意使用了英文中已有的 Pagoda 一词而不用汉字音译为 t'a。这是因为，在一切欧洲语言里，都采用这个词作为这种建筑物的名称，它已经被收入几乎所有欧洲语种的词典之中，作为中国塔的名称。这一事实，也可能反映了当西方人开始同中国接触时，八角塔在中国已多么流行。

这个时期塔的外部特征是日益逼真地模仿木构建筑。柱、枋、繁复的斗栱、带椽的檐、门、窗以及有栏杆的回廊等等，都在砖塔上表现出来。因此，这一时期的多层塔和密檐塔看起来与其早期的先型已有很大区别。

单层塔

在古拙时期风行一时的单层塔到了唐末以后已日趋罕见。仅存的 12 世纪以后的几例，平面都作方形。这种石室式的塔都筑于须弥座上，这种做法为唐代所未见。塔上的入口过去直通塔内，现在都做成了门，上有成排门钉及铺首。这类典型墓塔实例有河南嵩山少林寺中的普通禅师塔（公元 1121 年）[宋宣和三年]、行钧禅师塔（公元 926 年）[后唐天成元年]（图 64f）和西堂禅师塔（公元1157 年）[金正隆二年，原文年代有误]。

多层塔

当单层塔逐渐消失的时候，高塔也在发生变化。八角形的平面已成为常规，而方形的倒成了例外。原来在塔的内部被用来分隔和连通各层的木质楼板和楼梯已被砖石所取代。最初匠师们的胆子还小，他们把塔造得如同一座实心砖墩，里面只有狭窄的通道作为走廊和楼梯。但在艺高胆大之后，他们的这种砖砌建筑便日趋轻巧，各层的走廊越来越宽，最后竟成为一栋有一个砖砌塔心和与之半脱离的一圈外壳的建筑物，两者之间仅以发券或叠涩砌成的楼面相连。

这时期的多层塔又可分为两个子型，即仿木构式和无柱式，前者还可再分为北宗辽式和南宗宋式。

辽式仿木构式可以说是山西应县佛宫寺塔（公元 1056 年）[辽清宁二年]，即中国现存唯一的一座木塔（图 31）的砖构仿制品。在河北的典型实例有建于公元 1092 年[辽大安八年]的涿县双塔[即云居寺塔（俗称北塔）及智度寺塔（俗称南塔），原文及附图英文标题均有误，译文已改正]（图 67b，图 67c）和易县的千佛塔（图67a）。

此外，在热河[今内蒙古自治区]、辽宁等地也有若干遗例。它们对于木塔的忠实模仿一望可知，其比例上的唯一区别仅在于因材料所限而出檐较浅。

图 67

华北地区仿木构式
多层塔

67

Multi-storied
Pagodas of the Tim-
ber-frame Subtype,
North China

a. 河北易县千佛塔
（已毁于抗日战争
时期）

a. Ch'ien-fo T'a, I
Hsien, Hopei

b. 河北涿县云居寺
塔细部

b. Detail of North
Pagoda, Yun-chu
Ssu, Cho Hsien

c. 涿县云居寺塔

c. North Pagoda of
the Twin Pagodas of
Yun-chu Ssu, Cho
Hsien, Hopei, 1092

d. 河北正定广惠寺
华塔，约建于 1200
年（已毁）
d. Hua T'a
（Flowery Pagoda），
Kuang-hui Ssu,
Cheng-ting,
Hopei, ca.1200.
（Destroyed）

河北正定縣廣惠寺華塔平面圖
PLAN OF "FLOWERY PAGODA"
KUANG-HUI SSU, CHENG-TING, HOPEI
比例尺 5   0   10 M.

e. 华塔平面图
e. Hua T'a, plan

f. 河北房山云居寺
北塔
f. Pei T'a（North
Pagoda）, Yun-chu
Ssu, Fang Shan,
Hopei, Liao dynasty

　　河北正定县广惠寺华塔属于仿木构式，因其外形华丽而得名，并成为孤例（图67d，图67e）。这是一座八角形砖砌结构的塔，平面布局多变，组合奇妙；其外表模仿木构，塔的上部有一个装饰非常奇特的高大圆锥体；底层四角有四座六角形单层塔相连。这一特异布局可能是由房山县云居寺塔群变化而来。华塔建筑的确切年代已不可考，但从其所仿木构的特征来看，当是12世纪末或13世纪初的遗物。

　　北方仿木构式塔的另一个罕见的例子是房山云居寺北塔中央的主塔（图67f）。其四角有较早的唐代小石塔围绕。这座塔只有两层，显然是应县木塔类型的一个未完成的作品。塔顶为一巨型窣堵坡，有一个半球形的塔肚子和一个庞大圆锥形的"颈"［十三天］。塔的下面两层无疑是辽代所建，但上部建筑的年代可能稍晚。

南宋仿木构式塔在长江下游十分普遍。最早为浙江杭州灵隐寺的双石塔（图68a，图68b）和闸口火车站院内的白塔。它们通高约13英尺［约4米］，实际是塔形的经幢。其雕饰十分精美，无疑是对当时木构的忠实模仿。

南方型真正的塔的实例有江苏苏州［吴县］罗汉院的双塔（公元982年）［北宋太平兴国七年］（图68c—图68e）和虎丘塔（图68f，图68g）。与同期的北方型相比，它们在通体比例上显然较为纤细。这一特点由于塔顶有细长的金属刹，而塔下又没有高的须弥座而显得更为突出。从细部上看，这类塔的柱较短，但柱身卷杀微弱；斗栱较简单，第一跳偷心，又不用斜栱。在层数不多的叠涩檐内，用菱角牙子以象征椽头。因而出檐很浅，使塔在轮廓上迥然不同于北

图68

华南仿木构式多层塔

68

Multi-storied Pagodas of the Timber-frame Subtype, South China

a. 浙江杭州灵隐寺双石塔，建于公元960年

a. Twin Pagodas, Ling-yin Ssu, Hangchow, Chekiang, 960

b. 灵隐寺双石
塔细部
b. Detail of one
Twin Pagoda,
Ling-yin Ssu

c. 江苏吴县罗汉院
双塔渲染图
c. Twin Pagodas, Lo-
han Yuan, Soochow,
Kiangsu, 982,
rendering

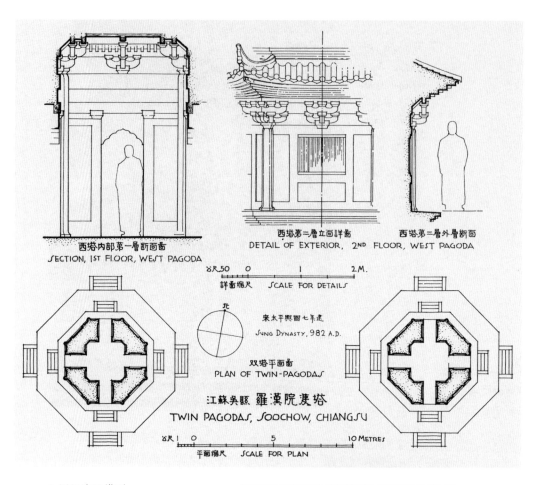

西塔内部第一層斷面畫
SECTION, 1ST FLOOR, WEST PAGODA

西塔第二層立面詳畫
DETAIL OF EXTERIOR, 2ND FLOOR, WEST PAGODA

西塔第二層外層斷面

δR 50　0　　　1　　2 M.
詳畫縮尺　SCALE FOR DETAILS

北

宋太平興國七年建
SUNG DYNASTY, 982 A.D.

雙塔平面畫
PLAN OF TWIN-PAGODAS

江蘇吳縣 羅漢院雙塔
TWIN PAGODAS, SOOCHOW, CHIANGSU

δR 1　0　　　　　5　　　　10 METRES
平面縮尺　SCALE FOR PLAN

d. 罗汉院双塔平
面、断面及详图
d. Twin Pagodas,
Lo-han Yan, plans,
section, and details

e. 罗汉院双塔之一
细部
e. Detail of one
Twin Pagoda, Lo-
han Yuan

156

f. 江苏吴县虎丘塔

f. Tiger Hill Pagoda,
Soochow, Kiangsu

g. 虎丘塔细部

g. Detail of Tiger
Hill Pagoda

方型。

福建省晋江［泉州市］的镇国塔和仁寿塔（公元 1228—1247年）［南宋理宗朝］是南方仿木构式的石塔，它们模仿木构形式极为忠实。由于这类塔几乎全系砖构，这两座石塔便成了难得的实例。

无柱式即立面不用柱的塔，主要是北宋形制。其遗例大多在河南、河北、山东诸省，其他地区较少见。这类塔的特点是墙上完全无倚柱，但常用斗栱。有些例子以砖砌叠涩出檐。斗栱不一定明显分朵，而常是排成一行，成为檐下一条明暗起伏相间的带。在大多数情况下，墙的上方都隐出阑额以承斗栱。

山东长清县灵岩寺辟支塔（图 69a）与河南修武县胜果寺塔，是这种砌有斗栱的无柱式塔的两个实例，均建于 11 世纪晚期。河北定县开元寺的料敌塔（公元 1001 年）[1] 则是叠涩出檐的一个精彩典型（图 69b）。这座塔的东北一侧已完全坍毁[2]，从而将其内部构造完全暴露在外，好似专为研究中国建筑的学生准备的一具断面模型（图 69c）。这类塔在西南地区也可见到，如四川省大足县报恩寺白塔（约公元 1155 年）［南宋绍兴年间］和泸县的塔等。

从宋代起，开始使用琉璃砖作为塔的面砖。河南开封市祐国寺（公元 1041 年）［宋庆历元年］[3] 的所谓"铁塔"就是华北地区著名的一个实例（图 69d，图 69e），塔为多层，转角处略微隐起倚柱，通体敷以琉璃砖，其色如铁，因而得此俗名。

其实，真正的铁塔在宋代也是有的。这种塔一般都很小，体形高瘦，这是其制作材料——铸铁——使然。这种微型塔与杭州石塔一样，实际上是一种塔形经幢，其遗例可见于湖北当阳玉泉寺和山东济宁。

密檐塔

唐亡之后，密檐塔仅见于辽、金统治地区，也是今天华北地区最常见的一种塔型。由于模仿木构，此时的这类塔与其古拙时期的先型相比，已大为改观。除少数罕见的例外，它们为八角形平面，

[1]
北宋咸平四年，此处原文所注系真宗下诏建塔年份，而塔建成年份为仁宗至和二年（公元 1055年）。——孙增蕃校注

[2]
此处原文及图注所述方向有误，译文根据作者所著《中国建筑史》改正。——孙增蕃校注

[3]
根据最近资料，该寺实建于 1049 年，即宋皇祐元年。——孙增蕃校注

图像中国建筑史

图 69

无柱式多层塔

69

Multi-storied Pago-
das of the Astylar

Subtype

a. 山东长清灵岩寺
辟支塔
a. P'i-chih T'a,
Ling-yen Ssu,
Ch'ang-ch'ing,
Shantung, late
eleventh century

b. 河北定县开元寺
料敌塔
b. Liao-ti T'a, K'ai-
yuan Ssu, Ting
Hsien, Hopei, 1001

c. 料敌塔西北侧
［应为东北侧］
c. Northwest side of
Liao-ti T'a

d. 河南开封祐国寺
铁塔
d. The "Iron
Pagoda," Yu-kuo
Ssu, K'aifeng,
Honan,1041

e. 祐国寺铁塔平
面图
e. "Iron
Pagoda," plan

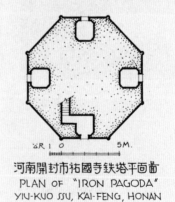

河南開封市祐國寺铁塔平面面
PLAN OF "IRON PAGODA"
YIU-KUO SSU, K'AI-FENG, HONAN

但结构上已是实心砌筑无法登临了。塔的下面无例外地都有一个很高的须弥座，再下又常筑有一个低而广的台基。主层转角处有倚柱，墙上隐起阑额和假门窗。各层出檐常以逼真的砖砌斗栱支承，但用叠涩出檐的也很常见。有时两种做法并用，遇此则只限于最下一层檐用斗栱。

这类塔最著名的一例是北京的天宁寺塔（图 70a）。在其须弥座之上还有一层莲瓣形平坐。塔上假门两侧有金刚像，假窗两侧则有菩萨像。塔建于 11 世纪，后世曾任意重修。

河北易县泰宁寺塔则是叠涩出檐的一个很好的实例［已于1960 年左右坍毁］。河北正定县临济寺青塔（公元 1185 年）［金大定二十五年］也属于这个类型，但规模较小；而赵县柏林寺（公元1228 年）［金正大五年］真际禅师塔也与此大同小异，唯每层檐下有一矮层楼面（图 70b）。这与一般制式不同，也可能是密檐塔与多层塔之间的一种折中类型。

这一时期的密檐塔有时仍保留古拙时期的某些传统，如采用方形平面。可能建于 13 世纪的辽宁朝阳县凤凰山大塔便是一例。河南洛阳市近郊的白马寺塔（图 70c），建于公元 1175 年［金大定十五年］。这也是一座方形塔，上有十三层檐。这种平面和全塔的比例显然仍有唐风。但这座塔是实心的，下面有须弥座，这两种做法又是前代所没有的。

另一个例子是建于公元 1102—1109 年［宋崇宁、大观间］的四川宜宾旧州坝白塔。如果仅看其外形，此塔直与唐塔无异，但其内部布局——连续的通道和梯楼环绕中心方室盘旋而上，则是这一时期的特征（图 70d）。

在西南各省，唐式方塔很多。特别是在云南省，直到清代还在建造这种形式的塔。

经　幢

经幢是一种独特的佛教纪念物，始见于唐代而盛于繁丽时期。

图 70

密檐塔

70

Multi-eaved Pagodas

a. 北京天宁寺塔

a. T'ien-ning
Ssu T'a, Peking,
eleventh century

b. 河北赵县柏林寺
真际禅师塔

b. Tomb Pagoda of
Chen-chi, Po-lin
Ssu, Chao Hsien,
Hopei, 1228

d. 四川宜宾白塔平
面及立面图
d. Pai T'a, I-pin,
Szechuan, ca.1102–
1109, plan and
elevation

c. 河南洛阳近郊白
马寺塔
c. Pai-ma Ssu
T'a, near Loyang,
Honan, 1175

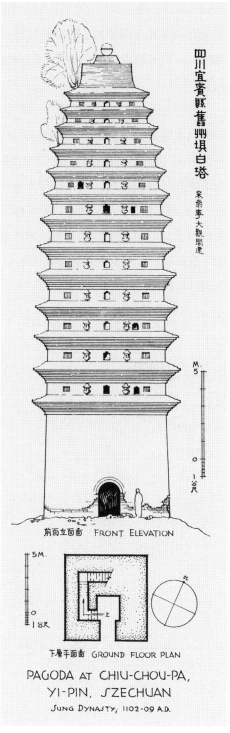

四川宜賓縣舊州壩白塔

宋崇寧大觀間建

M. 5

0

1 公尺

前面立面畫 FRONT ELEVATION

5M.

0

1 丈尺

北

下層平面畫 GROUND FLOOR PLAN

PAGODA AT CHIU-CHOU-PA,
YI-PIN, SZECHUAN
SUNG DYNASTY; 1102-09 A.D.

它又可称为经塔，视其与真塔区别大小而定。从建筑意味上说，这类纪念物彼此差异很大。最简单的是一根八角石柱，竖于一个须弥座上，柱端覆以伞盖；最复杂的则近似于一座小规模的、真正的塔。

建于公元857年［唐大中十一年］的山西五台山佛光寺大殿前的刻有施主宁公遇姓名的经幢（图71a），是简式经幢的一个典型。约建于12世纪的河北行唐县封崇寺经幢则是宋、金时代大量经幢中最具代表性的一种。现存经幢中最大的一座位于河北赵县（图71b），它雕饰精美，比例优雅，为宋初即11世纪所建。云南昆明地藏庵内建于13世纪的经幢也是一个有趣的实例。该幢看起来更近于一具石雕，而不是一个建筑物。在宋代灭亡之后，这种纪念性建筑似已不再流行。

图71
经幢
71

Dhanari Columns
a. 山西五台山佛光寺晚唐两经幢立面图
a. Dhanari column, Fo-kuang Ssu, 857
b. 河北赵县经幢
b. Dhanari column, Chao Hsien, Hopei, early Sung dynasty

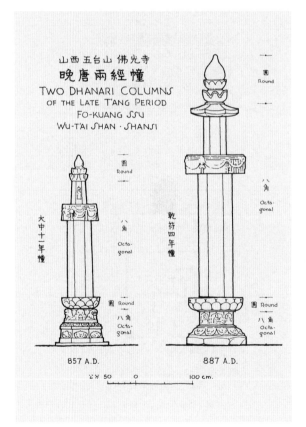

杂变时期

（约公元 1280—1912 年）

从 1279 年元朝建国开始，到 1912 年清朝覆亡，可称为塔的杂变时期。其第一个变化是，随着蒙古民族入主中原，喇嘛教也开始流传，于是瓶状塔（即西藏化的印度窣堵坡）突然大为流行。这种类型早在三个世纪前已在佛光寺塔上露其端倪；在金代，它作为僧人墓塔，也曾有过许多变种；最后到了元代，终于成为定制。

这一时期的第二个创新，是成形于明代的金刚宝座塔。其特点是筑五塔于一座高台之上。同样，这种形式也早有其先河，即 8 世纪初的河北房山县云居寺五塔（图 66k），但其后七百余年间，它却处于休眠状态，直到 15 世纪晚期才得复苏。尽管这种塔型在全国并不普遍，但现存实例已足可构成一种单独的类型。

明、清两代还有大量惯例形式的塔，此时建塔已不纯系事佛，而常常是为了风水。这种迷信认为自然界的因素——特别是地形和方向——会影响人们的命运，因而建塔以弥补风水上的缺陷。最常见的是文峰塔，即保佑科举考试交好运的塔。在南方诸省，此类塔甚多，大半筑于城南或城东南高处。

多层塔

自 1234 年金亡之后，密檐塔突然不再流行，而被多层塔所取代。在明代，这类塔的特点是塔身更趋修长，而各层更形低矮。在外形上，塔身中段不再凸出，较少卷杀，通体常呈直线形，收分僵直；屋檐的比例比原来木构小得多，出檐很浅，而斗栱纤细甚至取消，使屋檐沦为箍状。这类塔实例很多。陕西泾阳县的塔建于 16 世纪初，还葆有上代遗风中不少特点；而建于公元 1549 年［明嘉靖二十八年］的山西汾阳县灵严寺塔，则是一座典型的明代塔。山西太原永祚寺双塔建于公元 1595 年［明万历二十三年］（图 72a），其出檐较远，塔的外观由于檐下较深的阴影而比一般的明代塔显得明暗对比更强烈。

　　建于公元 1515 年［明正德十年］的山西赵城县［今洪洞县］广胜寺飞虹塔（图 72b）是一个特例。塔共十三层，塔身逐层收分甚骤，毫无卷杀，形成一座比例拙劣的八角锥体。尤为拙劣的是，在其底层周围，环有一圈过宽的木构回廊。塔的外面以黄、绿两色琉璃砖瓦赘饰，各层出檐则交替地以斗栱和莲瓣承托。在结构上，全塔实际上是一座实心砖墩，仅有一道楼梯盘旋而上，其结构颇有独到之处，全梯竟无一处供回转的平台（图 72c）。

图 72
杂变时期的多层塔
72
Period of Variety:
Multi-storied Pago-das
a. 山西太原永祚寺双塔
a. Twin Pagodas,
Yung-chao Ssu, Tai-yuan, Shansi, 1595

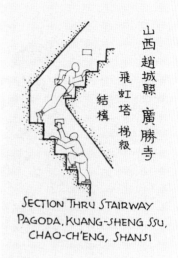

b. 山西赵城县广胜
寺飞虹塔

b. Fei-hung T'a,
Kuang-sheng Ssu,
Chao-ch'eng, Shan-
si, 1515

c. 飞虹塔梯级断
面图

c. Fei-hung T'a,
section through stair-
way

　　山西临汾县大云寺方塔（图 72d）大体也属这个类型，其外观也
同样不佳。此塔建于公元 1651 年［清顺治八年］，共五层，上更立
八角形顶一层，是刹的一个最新奇变体。底层内有一尊巨大的佛头
像，约高 20 英尺［6 米许］，直接置于地上。这种做法犹如放置此
像的塔的设计一样，全然不合常规。

　　在清代的多层塔中，还有几处值得一提。山西新绛县的塔属于
无柱式。虽然大体承袭前代传统，但卷杀过分，成为穗形。山西太
原市近郊晋祠奉圣寺塔，属于北方仿木构式，外形优美，有辽塔
之风（图 72e）。浙江金华市的北塔（图 72f）则是南方仿木构式的代
表作。

d. 山西临汾大云寺方塔

d. Square Pagoda, Ta-yun Ssu, Lin-fen, Shan-si, 1651

e. 山西太原晋祠奉圣寺塔

e. Feng-sheng Ssu T'a, Chin-tz'u, Taiyuan, Shansi

f. 浙江金华北塔

f. Pei T'a, Chin-hua, Chekiang, Ch'ing dynasty

密檐塔

自金灭亡之后，密檐塔遂不兴。北京的两对双塔，规模都较小，且虽原建于元，却在清代几乎全部经过重建。除此而外，目前所见唯一的"足尺"元代密檐塔是河南安阳市天宁寺塔（图 73c）[1]，这座塔有五重檐、五层。就其现状而言，像法国的许多哥特式大教堂的塔楼一样，显然是个未完成的作品。它与河北赵县柏林寺真际塔相似，每层檐下有一个非常矮的楼层。因此，严格地说，它并不是个密檐塔，而只是在总的外观上看来如此。塔顶的刹采用了清代典型的喇嘛塔。主层立面仿木构细部非常逼真地反映了当时的木构造。这座塔的平面布局也与其外观一样不同寻常，因为它不像辽、金的类似建筑那样是实心的。除在其底层内室四周筑有楼梯外，上部各层则为筒状结构，有如唐塔。这种平面在唐以后是很少见的。

尽管"足尺"的密檐塔在元代已不行时，却常有小型的建于墓地。典型实例如河北省邢台市的弘慈博化大士墓塔和虚照禅师墓塔（约公元 1290 年）[元朝初年]。后者为六角形，上面覆以一座半球形窣堵坡（图 73b）。地处西南边陲的云南省，受中原文化影响一般较迟，因此直到元代仍在建造密檐塔，但多依唐制，平面为方形，叠涩出檐。

明代留存的唯一一座密檐塔是北京八里庄慈寿寺塔（图 73a）。塔建于公元 1578 年［万历六年］，其造型显然曾深受附近的天宁寺塔（图 70a）的影响，整体比例近于辽制。但在细节上，它又明显具有明代晚期风格，如须弥座各层出入减少，塔身低矮，以及券窗、双层阑额和较小的斗栱等。

北京玉泉山塔，建于 18 世纪，是一座小型园林建筑。这是清代的一个有趣的创新，可说是多层与密檐的结合型，共三层，一、二层各有两重檐，第三层则有三重檐，看起来很别致。其平面虽是八角形，但并不等边，严格地说是削去四角的正方形。全塔以琉璃饰面，立面忠实地模仿了木构建筑。

图 73

杂变时期的密檐塔

73

Period of Variety:
Multi-eaved
Pagodas

a. 北京八里庄慈寿
寺塔［原文误为河
南安阳天宁寺塔］

a. T'ien-ning Ssu
T'a, Anyang, Hon-
an

b. 河北邢台虚照禅
师墓塔

b. Tomb Pagoda
of Hsu-chao,
Hsing-t'ai, Hopei,
ca.1290

c. 河南安阳天宁寺
塔［原文误为北京
八里庄慈寿寺塔,
原图亦改换］

c. Tz'u-shou Ssu
T'a, Pa-li-chuang,
Peking, 1578

喇嘛塔

如前所述，建于10世纪晚期的山西五台山佛光寺的半球形墓曾是喇嘛塔的先驱。其后，金代的一些墓塔也采用过这一形式；但直到元代，才正式成为雄伟建筑物的一种形制。此时，它又出现了一些新的形式，即在一个高台上筑一座瓶状建筑；台基一般为单层或双层须弥座，平面呈"亞"字形，台上有塔肚子和瓶颈状的"十三天"，再上则冠以宝盖。

北京妙应寺白塔可说是这类塔的鼻祖（图74a）。它是公元1271年［至元八年］根据元世祖忽必烈的敕令修建的。他有意拆毁了原有的一座辽塔，以便在其原址上另建此塔。与后来的同类塔相比，这座塔的比例肥短，塔肚子外侧轮廓线几乎垂直，而"十三天"则是一个截头圆锥体。

山西省五台山塔院寺塔（公元1577年）［明万历五年］，是明代瓶状塔的一个显著例子（图74b）。它的塔肚底部略向内收，而"十三天"的上部却增大了。总的看来，显得比上述白塔苗条一些。

山西代县善果寺塔也建于明代，但确实年代已无考。这座塔的须弥座呈圆形，与塔身的比例比通常大得多，造型简洁有力，上层须弥座的束腰收缩较多。它的塔肚轮廓柔和，"十三天"的底部又有一圈须弥座。塔形通观稳重雅致，可以说是中国现存瓶状塔中比例最好的一座。

此后，喇嘛塔在比例上日趋苗条，特别是其"十三天"。这种趋势的两个实例是北京北海公园内的永安寺白塔（公元1651年）［清顺治八年］（图74c）和距北京西山的围场园子——香山静宜园不远的法海寺遗址拱门石台上的喇嘛塔（公元1660年）［清顺治十七年］。两者虽然规模悬殊，通体比例却几无二致。它们的须弥座都简化为一层，座上塔肚以下的部分加高了，以代替过去的凹凸线道，而在法海寺则成为另一阶级形的座，与下面的须弥座形状略似。塔肚上设龛，现在已成为定制。"十三天"收分极少，几乎成为圆柱形，与塔肚相较，比例上甚显细瘦。

a. 北京妙应寺白塔
a. Pai T'a, Miao-
ying Ssu, Peking,
1271

b. 山西五台山塔院
寺塔
b. T'a-yuan Ssu
T'a, Wu-t'ai Shan,
Shansi, 1577

图 74
杂变时期的喇嘛塔
　74
Period of Variety:
Lamaist Stupas

辽宁沈阳附近地区也有几座同类型的塔，都属清初遗物。其特点是底座和塔肚特别宽大。这类塔的许多变型常可见于寺庙院内，多由青铜铸成，体形很小。以山西五台山显庆寺大殿前的诸塔为其代表。

金刚宝座塔

明代在筑塔方面的重要贡献之一，是确立了金刚宝座塔这种塔型。其特征是五塔同筑于一个台基之上。其先型曾见于河北房山县云居寺诸塔（公元711—727年）[1]（图66k）。其后就是山东历城县柳埠村[2]的九塔寺塔，建于唐中叶，约公元770年，是这类塔中一个更加华丽的实例，竟集九座小型的密檐方塔于一座八角形单层塔之上。在金代，这种构思又演变出河北正定县华塔这种怪异的形式（图67d）。然而，作为一种塔型的确立，以云南昆明市近郊妙湛寺塔为其标志，则是明天顺年间（公元1457—1464年）的事了。这是一个台基上的五座喇嘛塔，其"十三天"修长，略作卷杀，为

图 75
杂变时期的五合
一塔
75
Period of Varitty:
Five-pagoda
Clusters

a. 北京大正觉寺
（五塔寺）塔
a. Wu T'a Ssu
（Five-pagoda
Temple）, Cheng-
chueh Ssu, Peking,
1473

b. 北京碧云寺金刚
宝座塔
b. Chin-kang Pao-
tso T'a, Pi-yum Ssu,
Peking, 1747

喇嘛塔中所仅见；而在塔肚上设龛，在当时也极少有。台基下有十字形券道，但不通塔上。

这类塔中最重要的一例是北京西郊大正觉寺［俗称五塔寺］的金刚宝座塔（图 75a）。这座塔建于公元 1473 年［明成化九年］，台基分为五层，周围各有出檐，看去如同西藏寺庙。南面有拱门，内有梯级通往台顶，台上有密檐方塔五座，正前方小亭一座，成为梯级的出口处。

北京西山［香山］碧云寺金刚宝座塔是另一重要实例（图 75b，图 75c），在这组建于公元 1747 年［清乾隆十二年］的塔中，五塔寺的那种布局更趋复杂。在五座密檐方塔之前，又增加了两座瓶形塔，两塔间稍后处的小亭本身又成为小型台基，上面再重复了五塔的布置。整组塔群耸立于两层毛石高台之上。

北京城北黄寺的群塔，规模较前述各塔要小得多。中央的喇嘛塔形状奇特，角上的四塔则为八角多层式。塔座和下面的台基都很矮。塔前还有一座牌楼。

c. 金刚宝座塔平面及立面图
c. Pi-yun Ssu, plan and elevation

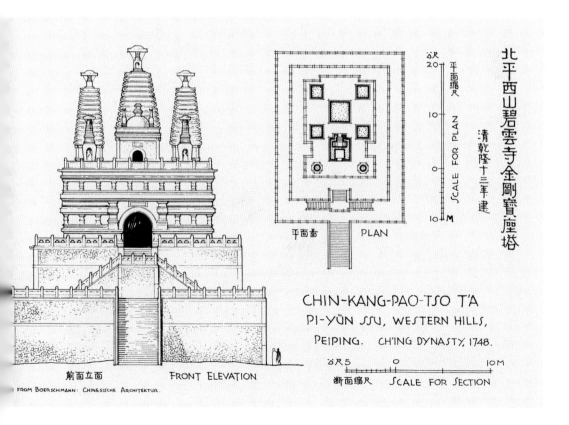

北平西山碧雲寺金剛寶座塔

清乾隆十三年建

平面圖　PLAN

平面縮尺　SCALE FOR PLAN

前面立面　FRONT ELEVATION

CHIN-KANG-PAO-TSO TA
PI-YÜN SSU, WESTERN HILLS,
PEIPING. CH'ING DYNASTY, 1748.

断面縮尺　SCALE FOR SECTION

FROM BOERSCHMANN: CHINESISCHE ARCHITEKTUR.

其他砖石建筑

中国的匠师对于以砖石作为常用的主要结构材料这一点，一直是不甚了解的。砖石或被用于与日常生活关系不太密切的地方，如城墙、围墙、桥涵、城门、陵墓等，或被用于次要的地方，如木构房屋的非承重墙、窗台以下的槛墙等等。所以，砖石结构在中国建筑中与在欧洲建筑中的地位是无法相提并论的。

陵 墓

现存最古老的券顶结构是汉代的砖墓，其数量极大。这种地下构筑物从未以建筑手法来修建，因而从建筑学的角度来看，其意义不大。在地面上，墓前通常有一条大道，入口处有一对阙，然后是石人、石兽，最后是陵前的享殿。在这里，只有阙和享殿具有建筑上的意义，而那些石雕，对于学雕塑的学生比学建筑的学生更为重要。至于六朝和唐代的陵墓，由于只剩下了石雕，我们就更不感兴趣了。

在四川宜宾和南溪附近，曾偶然发现了几座 12 世纪的坟墓。它们显示出南宋时期在坟墓中采用了很高程度的建筑处理手法。这些用琢石砌筑的墓室虽然不大，许多地方却竭力模仿当时的木构建筑。正对入口的一端无例外地是两扇半掩着的门，门后半露出一个女子形象（图 76a）。这种类型的墓在中国其他地区尚未见到，它是否仅为本地区所特有，尚待研究。

明、清两代陵墓，遗例很多。其中最重要的是皇帝的陵墓。这种陵墓的地上建筑主要是陵前的一座座殿堂，所以，把这样的建筑群称为"陵寝"，倒也恰如其分。其唯一不同之处，是在地宫的入口处筑有方城一座，城上建有明楼。位于河北［北京市］昌平县的明永乐长陵是河北省内昌平县［明十三陵］、易县［清西陵］和兴隆县［清东陵］三处皇陵中最宏伟的一座（图 76b，图 76c）。

易县清嘉庆昌陵（公元 1820 年）［嘉庆二十五年］是地下宫殿的典型实例［北京明十三陵定陵发掘于本书成稿后十年。——译注］。

图 76

陵墓

76

Tombs

a. 四川宜宾无名墓，约建于 1170 年

a. Unidentified tomb, I-pin, Szechuan, ca. 1170

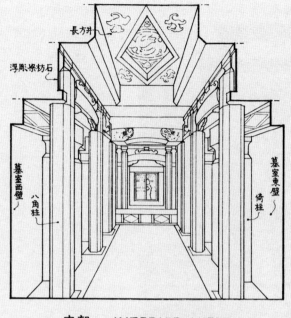

内部　INTERIOR VIEW

平面　PLAN

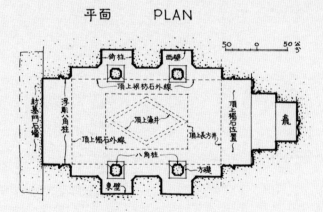

AN UNIDENTIFIED TOMB C.1170

I-PIN　SZECHUAN

b. 河北明长陵方城
及明楼

b. Tomb of Ming
emperor Yung-
lo, Ch'ang-
p'ing, Hopei,
1415, showing
"radiant tower"
(*ming-lou*) set on
"square bastion"
(*fang-ch'eng*)

c. 明长陵总平面图

c. Yung-lo's tomb,
plan

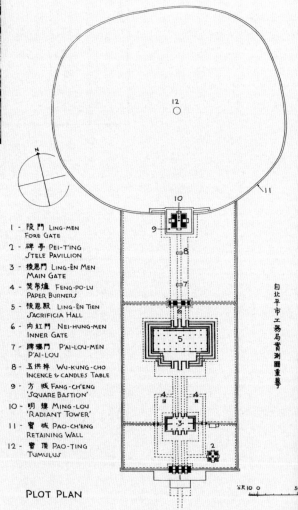

1 - 陵門 LING-MEN
    FORE GATE
2 - 碑亭 PEI-T'ING
    STELE PAVILLION
3 - 棱恩門 LING-ÊN MEN
    MAIN GATE
4 - 焚帛爐 FENG-PO-LU
    PAPER BURNERS
5 - 棱恩殿 LING-ÊN TIEN
    SACRIFICIA HALL
6 - 内紅門 NEI-HUNG-MEN
    INNER GATE
7 - 牌樓門 P'AI-LOU-MEN
    P'AI-LOU
8 - 五供棹 WU-KUNG-CHO
    INCENCE & CANDLES TABLE
9 - 方城 FANG-CH'ENG
    'SQUARE BASTION'
10 - 明樓 MING-LOU
    'RADIANT TOWER'
11 - 寶城 PAO-CH'ENG
    RETAINING WALL
12 - 寶頂 PAO-TING
    TUMULUS

PLOT PLAN

CH'ANG-LING · TOMB OF EMPEROR YUNG
CH'ANG-P'ING · HOPEI ·· MING DYNASTY · 1409-24
REDRAWN AFTER PLAN BY THE BUREAU OF CONSTRUCTION · MUNICIPAL GOVERNMENT OF

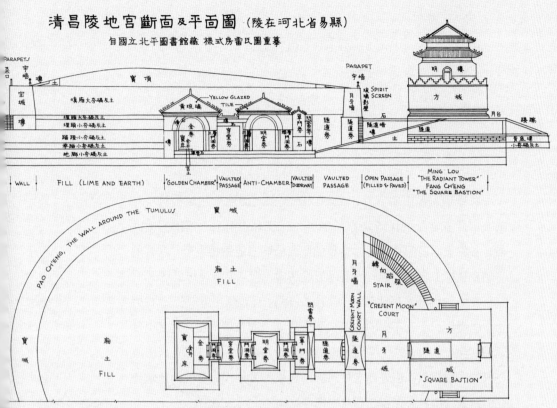

清昌陵地宫断面及平面图 (陵在河北省易县)
自国立北平图书馆藏 样式房雷氏图重摹

CH'ANG LING, TOMB OF EMPEROR CHIA-CH'ING, 1796-1820, CH'ING DYNASTY
PLAN AND SECTION OF SUBTERRANEAN TOMB CHAMBERS, REDRAWN AFTER ORIGINAL DRAWINGS
THE LEI FAMILY, HEREDITARY OFFICIAL ARCHITECTURAL DESIGNERS. (COLLECTION, NATIONAL PEIPING LIBRARY).

d. 清昌陵地宫断面
及平面图
d. Tomb of Ch'ing
emperor Chia-
ch'ing, I Hsien,
Hopei, 1820, site
plan and elevation

在方城下面，有一条隧道通往一连串的罩门、明堂和穿堂，最后到达金券，即安放皇帝棺椁之处。这些券顶地宫都以雕花白石砌筑。一般还以黄琉璃瓦盖顶，与地上建筑无异，只不过上面覆以三合夯土，形成宝顶而已。有些陵墓未用瓦顶，多系承前帝遗旨，以示其节俭之德。世袭了数百年的清宫样式房雷氏家中所藏的档案，更确切地说是故纸堆中的图样，提供了有关这些地宫的宝贵资料（图76d）。

券顶建筑

在中国，除山西省外，全部以砖石砌成的地上建筑是极少见的。山西省常见的民居是券顶窑居，一般有三至七个筒形券顶并列，各券之间以门孔相通，券顶的截面成椭圆或抛物线形，敞开的一端在槛墙上装窗。券肩填土以形成平顶，可从室外梯级攀登。

寺庙的大殿有时也用券顶结构，称为无梁殿。建于公元1597年［明万历二十五年］的山西太原永祚寺，即双塔所在，是其中最好的实例（图77a，图77b）。这座殿的券顶是纵向的，门窗孔与山西一般的民居相似，但外面有柱、额、斗栱等的处理。类似的建筑物在山西五台山显通寺和江苏苏州都可见到（图77c，图77d）。

在明代中叶以前，还没有将砖筑券顶建筑的外表模仿成一般木构殿堂的做法，虽然在砖塔上已很常见。这种做法与欧洲文艺复兴时期的建筑采用古希腊、罗马时代古典柱式的做法相仿。值得一提的是，当公元1587年［明万历十五年］利玛窦到达南京，耶稣会开始对中国文化发生影响时，山西原有的券顶窑居已经为这种形式的建筑的发展提供了一个很充分的基础。耶稣会传教士到达中国，与无梁殿在中国的出现恰好同时，这也许并非巧合。

北京郊区还有几座清代的无梁殿（图77e），都比山西所见明代遗构要大得多。它们外观无柱，仿佛藏在厚重的墙内，而只以琉璃砖砌出柱头上的额枋和斗栱。目前还不了解，为什么这种宏伟壮观的建筑未能普及。

图 77

77

无梁殿

Vaulted "beamless halls" (*wu-liang tien*)

a. 山西太原永祚寺
a. Yung-chao Ssu, T'aiyuan, Shansi, 1597

b. 永祚寺砖殿平面图
b. Yung-chao Ssu, plan

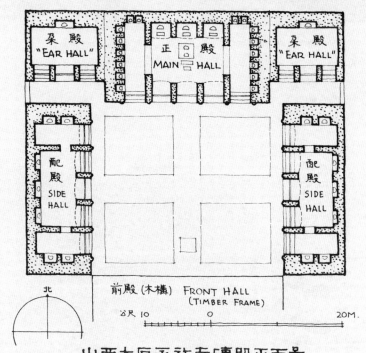

c. 江苏苏州开元寺
无梁殿

c. K'ai-yuan Ssu,
Soochow, Kiangsi

d. 开元寺无梁殿内
藻井

d. K'ai-yuan Ssu,
interior detail of
vaulting

e. 北京西山无梁
殿，建于18世纪

e. Beamless hall,
Western Hills,
Peking, eighteenth
century

桥

中国最古老的桥是木桥，河面宽时则用浮桥。文献中关于拱桥的记载最早见于 4 世纪。

中国现存最古老的拱桥是河北赵县近郊的安济桥，俗称"大石桥"（图 78a，图 78b）。这是一座单孔［空撞券］桥，在弧形主券两肩又各有两个小撞券。从露出河床上的两点算，其跨度为 115 英尺［35 米］，但加上埋入两岸土中部分，其净跨应大于此数。当年我们曾在桥墩处发掘，试图寻找其起拱点，但由于河床下 2 米左右即见水而未能成功［后经实测，起拱点之间的净跨为 37.47米。——译注］。

这座桥是隋朝（公元 581—618 年）建筑大匠李春的作品。主券由二十八道独立石券并列组成。这位匠师显然深知各道券有向外离散的危险，所以将桥面部分造得略窄于下端，从而使各道石券都略向内倾，以克服其离心倾向。然而，他的预见和智慧未能完全经受住时间和自然的考验，西侧的五券终于在 16 世纪坍毁［不久即修复］，而东侧的三券也在 18 世纪坍倒。

赵县西门外还有一座式样相同但规模较小的桥，人称"小石桥"［永通桥］（图 78c）。是公元 1190—1195 年［金明昌年间］由匠师褒钱而所建[1]，显然是模仿"大石桥"，但长度只略过后者的一半。这座桥的栏杆修于公元 1507 年［明正德二年］，它与早期的木构栏杆十分接近，是从早期仿木构式演化为明清流行式的一个过渡，因此是一个值得注意的实例。栏杆下部华版的浮雕图案亦很

［1］
建造永通桥的匠师姓名已失传，原文及图 78c 注字均误。按《畿辅通志》中有"赵人袁钱而建"一语，意为赵县人民集资兴建，该图的绘制者将"袁"误看作"褒"，并误为人名。——孙增蕃校注

有趣。

　　同宫室一样，清代的桥在设计上也标准化了（图78d）。北京附近有许多这种官式桥，其中最著名的就是卢沟桥，西方称为"马可·波罗桥"。这座桥原建于12世纪末，即金明昌年间，后毁于大水。现存的这座十一孔、全长约1000英尺［实测266.5米］的桥是18世纪重修的。这就是1937年日本军队在一次"演习"中对中国守军发动突然袭击——即"卢沟桥事变"——的历史性地点，这次事变导致了全国性的抗日战争（图78e）。

　　南方各地的拱桥在结构上一般比官式桥要轻巧些。如浙江金华县的十三孔桥就是一个杰出的范例（图78f）。南方还常可见到以石头作桥墩，上架以木梁，再铺设路面的桥，云南富民县桥可作为代表。陕西西安附近浐河、灞河上的桥是以石鼓砌桥墩，再以木头架作桥板（图78g，图78h）。在福建省，常可见到用大石板铺成的桥。在四川（如西康）、贵州、云南诸省，还广泛使用了悬索桥（图78i，图78j）。

图78
桥
78
Bridges
a. 河北赵县安济桥
（大石桥）
a. An-chi Ch'iao
（Great Stone
Bridge）, near
Chao Hsien, Hopei,
Sui, 581–618

　　　　　　　　图像中国建筑史

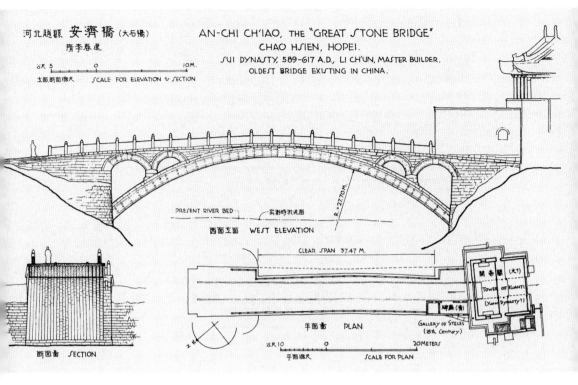

河北赵县 **安济橋** (大石橋)
隋李春建

AN-CHI CH'IAO, THE "GREAT STONE BRIDGE"
CHAO HSIEN, HOPEI.
SUI DYNASTY, 589-617 A.D., LI CH'UN, MASTER BUILDER.
OLDEST BRIDGE EXISTING IN CHINA.

营造尺 5   0   10M.
立面断面缩尺   SCALE FOR ELEVATION & SECTION

R = 27.70 M.

PRESENT RIVER BED   —实测时淤泥面

西面立面   WEST ELEVATION

CLEAR SPAN 37.47 M.

开帝阁 (大1)
TOWER OF KUANTI
(YUAN DYNASTY 1)

断面 (青)
碑廊 GALLERY OF STELES
(18th Century)

平面图   PLAN

营造尺 10   0   20 METERS
平面缩尺   SCALE FOR PLAN

断面图 SECTION

b. 安济桥立面、断面及平面图
b. An-chi Ch'iao, plan, elevation, and section

c. 河北赵县永通桥（小石桥）立面图
c. Yung-t'ung Ch'iao（Little Stone
Bridge）, Chao Hsien, Hopei, late twelfth
century elevation

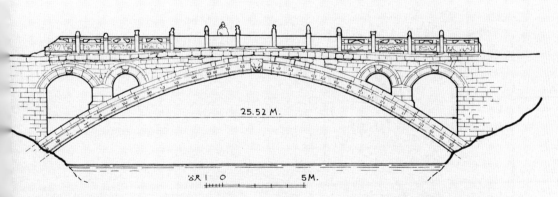

25.52 M.

营造尺 1 0   5M.

河北赵县 **永通橋**
俗呼小石橋 金明昌间
襄铵而建

YUNG-T'UNG CH'IAO OR LITTLE STONE BRIDGE
CHAO HSIEN, HOPEI, MING-CH'ANG PERIOD, 1190-95, CHIN DYNASTY.

清官式三孔石橋做法要略 RULES FOR DESIGN AND CONSTRUCTION OF A THREE-ARCHED BRIDGE IN OFFICIAL CH'ING STYLE

Brick Back-up

背後石　扣鼻石　牙子石
如意石下灰土

撞券頭　鵝鳳瓦鎮下土灰土
Lime Concrete
金剛牆肩後浪灰土

撞券背後料撞面的灰牆

次孔

土

券山石
長14分

撞券背後料撞面的灰牆

仰天石　內券石
分水金剛牆
券臉石　分水石　雁翅

河身泊岸
背道牆灰土

2×4分
4×15分

HALF LONGITUDINAL SECTION 縱斷面

17分　10分　19分　10分　17分　15分

Width of River (Distance between Enbankments)=103分 河口寬 (由泊岸間距離)

146分

孔數 NUMBER OF ARCHES	河口寬度分作 DIST. BT'N ENBANK.MENT DIVIDED INTO	孔寬 SPAN OF ARCHES (IN FEN) 中中方石 COUNTING FROM CENTRAL ARCH							
		中 C	I	II	III	IV	V	VI	VII	VIII
3	103分 fen	19	19	—	—	—	—	—	—
5	153分	19	17	15	—	—	—	—	—
7	199分	19	17	15	13	—	—	—	—
9	251分	19	17½	16	14½	13	—	—	—
11	294分	19	17½	16	14½	13	11½	—	—
13	355分	19	18	17	16	15	14	13	—
15	399分	19	18	17	16	15	14	13	12
17	441分 fen	19	18	17	16	15	14	13	12	11

左面 HALF ELEVATION

59.6分
33.2分
13.08分　6.64分　13.08分

柱子欄板
金邊
象眼
雁門石
仰天石
橋面鵝鳳瓦

分水金剛牆一律寬 10分；高 6分。
All Piers: Width=10 fen; Height=6 fen.
雁翅一律出 15分；高 6分。
Abutments: Width=15 fen; Height=6 fen.
關於其他細節參閱
For Further Details, cf.
王璧文：清官式石橋做法

尺　分
尺度縮尺

分度縮尺

管造尺縮尺

SCALE IN CH'IH (CHINESE FOOT = 32 cm)

SCALE IN FEN

雁翅上泊岸　券牆　內券石　臺帽臺　河身泊岸
Up-stream Lime Concrete Pavement　雁翅　分水金剛牆　分水尖　Down-stream Lime Concrete Pavement
迎水灰土　順水灰土

河身泊岸下地丁　聚椿下丁　牙丁
Piles under Embankment　Guard Piles　Piles under Piers & River-bed Pavement　牙丁 Guard Piles

CROSS SECTION 橫斷面

Up-stream Lime-concrete Pavement

Guard Piles　on Edge

河身泊岸灰土　迎水外牙子　15分
Brick Back-up

迎水外牙子　Water Breaker
迎水礓磋板
Flat
雁翅

Pier
分水尖

橋面礓磋板後浪牆
兩邊金剛牆
河身泊岸灰土

37.4分
橋面礓磋板後浪牆

券牆

Lime Concrete Back-up

21.2分

河身泊岸灰土
背道磚灰土
背道牆

5分　5分

河口寬度 103分

鳳凰臺
柱子
橋心　39.6分
牙子石
如意石

57.16分

分水礓磋板
Abutment
分水尖外牙子石
Down-stream Lime-Concrete Pavement
順水灰土寬30分
Embankment

River-bed Pavement

券牆背後浪牆

River-bed Pavement

Guard Piles

PLAN · PIER LEVEL 金剛牆平面　　　橋面平面 **PLAN · ROADWAY LEVEL**

d. 清官式三孔石桥
做法要略（左页）
d. Ch'ing rules for
constructing a three-
arched bridge

e. 北京卢沟桥
e. Lu-kou Ch'iao
（Marco Polo
Bridge）, Peking,
eighteenth century
f. 浙江金华十三孔
桥，建于 1694 年
f. Thirteen-arched
bridge, Chin-hua,
Chekiang, 1694

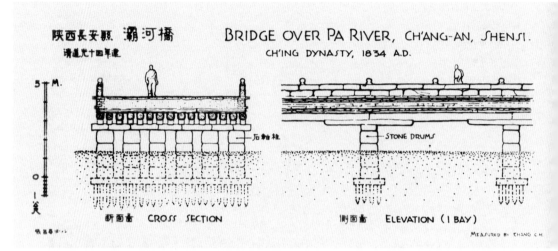

陕西长安县 灞河桥
清道光十四年建

BRIDGE OVER PA RIVER, CH'ANG-AN, SHENSI.
CH'ING DYNASTY, 1834 A.D.

5 — M.

0 —

1 尺

断面圖 CROSS SECTION

石軸柱

侧面圖 ELEVATION (1 BAY)

STONE DRUMS

MEASURED BY CHANG C.H.

g. 灞河桥断面及侧
面图
g. Bridge over Pa
River, Sian, Shensi,
1834, elevation and
section

h. 灞河桥细部
h. Detail of Pa River
bridge

192

i. 四川灌县竹索桥，
建于 1803 年
i. Bamboo
suspension bridge,
Kuan Hsien,
Szechuan, 1803

j. 竹索桥断面、立
面及平面图
j. Kuan Hsien
suspension bridge,
elevation, plan, and
section

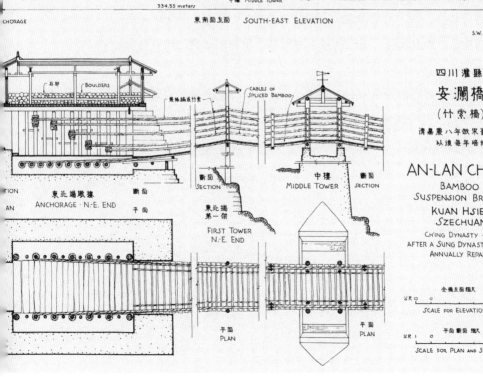

台

　　早在殷代，已常有为了娱乐目的而筑台的做法。除史籍中有大量关于筑台的记载外，在华北地区，至今仍有许多古台遗迹。其中较著名的，有今河北易县附近战国时燕下都（公元前 3 世纪末）遗址的十几座台。但都只剩下一些 20—30 英尺［7—10 米］左右的土墩，其原貌已不可考了。

　　用于宗教目的的台称作坛。北京天坛的圜丘就是其中最为壮观的一座。天坛是每年农历元旦黎明时皇帝祭天的地方，创建于公元 1420 年［明永乐十八年］，但曾于 1754 年［清乾隆十九年］大部重修。圜丘是以白石砌筑的一座圆坛，共有三层，逐层缩小，各有栏杆环绕，四方有台阶通向坛上（图 79a）。其他地方的坛，如北京的地坛、先农坛等，则仅仅是一些低矮、单调的平台，没有什么装饰。

　　河南登封县附近告成镇的测景台（图 79b），从建筑学的角度看意义不大，但可能使研究古天文学的学生大感兴趣。这是元代郭守敬所筑的九台之一，用以在每年冬至和夏至两天观测太阳的高度角。

图 79

台

79

Terraces

a. 北京天坛圜丘

a. Altar of Heaven
（Yuan-ch'iu），
at Temple of
Heaven,Peking,
1420, repaired 1754

b. 河南登封县告成
镇测景台平面及透
视图

b. Ts'e-ching T'ai,
a Yuan observatory,
Kao-ch'eng Chen,
near Teng-feng,
Sung Shan, Honan,
ca.1300

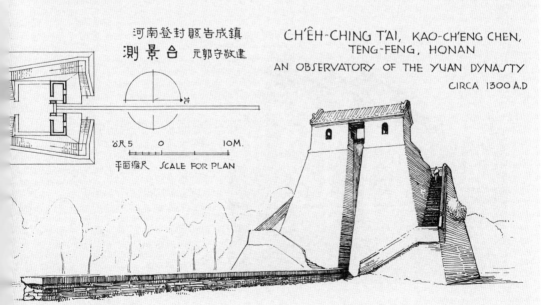

河南登封县告成镇
测景台 元郭守敬建

CH'ÊH-CHING TAI, KAO-CH'ENG CHEN,
TENG-FENG, HONAN
AN OBSERVATORY OF THE YUAN DYNASTY
CIRCA 1300 A.D

营尺 5　0　　　10M.
平面缩尺 SCALE FOR PLAN

MEASURED BY LIU T.T.

牌　楼

　　中国所特有的牌楼，是用来使入口处壮观的一种建筑物，如同汉代的阙一样。作为标志性的独立大门，它可能曾受到印度的影响，而不仅与著名的印度桑溪窣堵坡入口（建于公元前 25 年）偶然相似而已。

　　现存最古的牌楼可能要推河北正定县隆兴寺牌楼，它大约建于宋代，但上部在后世修缮时原貌已大改。在《营造法式》中，把与此类似的大门称为乌头门。在唐代文献中，也常见到这个名词。然而，直到明代，这类建筑物才广为流行。

　　这类建筑物中最庄严的一座是河北昌平县［今北京］明十三陵入口处的白石牌楼（公元 1540 年）［嘉靖十九年］（图 80a）。清代皇

图 80

牌楼

80

P'ai-lou Gateways

a. 河北昌平明十三陵入口处白石牌楼

a. Marble *p'ai-lou*, at entrance to Ming Tombs, Ch'ang-p'ing, Hopei, 1540

陵也有类似的牌楼，但规模较小。全国其他地方还有不计其数的石牌楼，其形制各异。四川广汉县的五座牌楼，就其单座形式而言是典型的，但成组布置却很少见，实为壮观（图 80b）。

在北京，木构牌楼很多。[成贤]街上的一座以及颐和园湖前的一座是其中的两个典型（图 80c，图 80d）。此外，北京还有不少把砖砌券门做成牌楼形式，并饰以琉璃的（图 80e）。

b. 四川广汉附近的五座牌楼
b. Five related *p'ai-lou*, near Kuang-han, Szechuan

c. 北京街道上的牌楼
c. Street *p'ai-lou*, Peking

d. 北京颐和园湖前牌楼

d. Lakefront *p'ai-lou*, Summer Palace, Peking

e. 北京国子监琉璃牌楼

e. Glazed terra-cotta *p'ai-lou*, Kuo-tzu Chien, Peking

梁思成传略

费慰梅

长久以来，对于我们西方人来说，中国的传统建筑总因其富于异国情趣而令人神往。那些佛塔庙宇中的翼展屋顶，宫殿宅第中的格子窗棂，庭园里的月门和拱桥，无不使 18 世纪初的欧洲设计家们为之倾倒，以致创造了一种专门模仿中国装饰的艺术风格，即所谓 Chinoiserie。他们在壁纸的花纹、瓷器的彩绘、家具的装饰上，到处模仿中国建筑的图案，还在阔人住宅的庭院里修了许多显然是仿中国式样的东西。这种上流阶层的时尚于 1763 年在英国可谓登峰造极——"克欧花园"（Kew Gardens）[1]中建起了一座中国塔；而且此风始终不衰。

在中国，工匠们千百年来发展出这些建筑特征，则是为了适应人们的日常之需，从避风遮雨直到奉侍神明或宣示帝王之威。奇怪的是，建筑却始终被鄙薄为匠作之事而引不起知识界的兴趣去对它做学术研究。直到 20 世纪，中国人才开始从事于本国建筑史的研究工作，而其先驱者就是梁思成。

梁思成（1901—1972）的家学和教育，使他成为中国第一代建筑史学家领导者的最适合人选。他是著名的学者和改革家梁启超的长子。他热爱父亲并深受其影响，将父亲关于中国的伟大传统及其前程的教诲铭记在心。他的身材不高，却有着镇于

---

〔1〕 即英国皇家植物园的邱园。——编者注

观察、长于探索、细致认真和审美敏锐的天资，喜爱绘画并工音律。梁思成在父亲
被迫流亡日本时出生于东京，在北京长大。在这里，他受到两个方面的早期教育，
后来的事实表明，它们对于他未来的成就是极其重要的。首先，是在他父亲指导下
的传统教育，也就是中国古文的修养，这对于日后他研读古代文献、辨识碑刻铭文
等都是不可少的；其次，是在清华学校中学到的扎实的英语、西方自然科学和人文
学科知识，这些课程是专为准备出国留学的学生开设的。他和他的同学们都属于中
国知识分子中杰出的一代，具有两种语言和两种文化的深厚修养，在沟通中西文化
方面成绩卓著。

　　梁思成以建筑为其终身事业也有其偶然原因。这个选择是一位后来同他结为夫
妻的姑娘向他建议的。这个姑娘名叫林徽因，是学者、外交家和名诗人林长民之女。
1920 年［此处英文原文有误］，林长民奉派赴英，年甫 16 岁的林徽因被携同行，她
非常聪颖、敏捷和美丽，当时就已显示出对人具有一种不可抗拒的吸引力，后来这
成了她一生的特色。她继承了父亲的诗才，但同样爱好其他艺术，特别是戏剧和绘

画。她考入了一所英国女子中学，迅速地增长了英语知识，在会话和写作方面达到了非常流利的程度。从一个以设计房子为游戏的英国同学那里，她获知了建筑师这种职业。这种将日常艺术创造与直接实用价值融为一体的工作深深地吸引了她，认定这正是她自己想要从事的职业。回国以后，她很快就使梁思成也下了同样的决心。

他们决定同到美国宾夕法尼亚大学建筑系学习。这个系的领导者是著名的保罗·克雷特，一位出身于巴黎美术学院的建筑家。梁思成的入学由于1923年5月在北京的一次摩托车车祸中左腿骨折而被推迟到1924年秋季。这条伤腿后来始终没有完全复原，以致落下了左腿略跛的残疾。年轻时，梁思成健壮、好动，这个残疾对他损害不大；但是后来却影响了他的脊椎，常使他疼痛难忍。

宾大的课程继承了巴黎美术学院的传统，旨在培养开业建筑师，但同样适合于培养建筑史学家。它要求学生钻研古希腊、罗马的古典建筑柱式以及欧洲中世纪和文艺复兴时期的著名建筑。系里常以绘制古代遗址的复原图或为某未完成的大教堂作设计图为题，举行作业评比，以测验学生的能力。对学生的一项基本要求是绘制整洁、美观的建筑渲染图，包括书写。梁思成在这方面成绩突出。在他回国以后，对自己的年轻助手和学生也提出了同样高标准的要求。

梁思成在宾大二年级时，父亲从北京寄来的一本书决定了他后来一生的道路。这是公元1103年由宋朝一位有才华的官员辑成的一部宋代建筑指南——《营造法式》，其中使用了生僻的宋代建筑术语。这部书已失传了数百年，直到不久前它的一个抄本才被发现并重印。梁思成立即着手研读它，然而如他后来所承认的那样，却大半没有读懂。在此之前，他很少想到中国建筑史的问题，但从此以后，他便下了决心，非把这部难解的重要著作弄明白不可。

林徽因也在1924年的秋季来到了宾大，却发现建筑系不收女生。她只好进入该校的美术学院，设法选修建筑系的课程。事实上，1926年她就被聘为"建筑设计课兼任助教"，次年又被提升为"兼任讲师"。1927年6月，在同一个毕业典礼上，她以优异成绩获得美术学士学位，而梁思成也以类似的荣誉获得建筑学硕士学位。

在费城克雷特的建筑事务所里一同工作了一个暑假之后，他们两人暂时分手，分别到不同的学校深造。由于一直对戏剧感兴趣，林徽因来到耶鲁大学乔治·贝克著名的工作室里学习舞台设计；梁思成则转入哈佛大学，以研究西方学者关于中国艺术和建筑的著作。

就在这一时期，正当梁思成二十多岁的时候，第一批专谈中国建筑的比较严肃的著作在西方问世了。1923 年和 1925 年，德国人厄恩斯特·伯希曼出版了两卷中国各种类型建筑的照片集。1924 年和 1926 年，瑞士一位艺术史专家喜龙仁发表了两篇研究北京的城墙、城门以及宫殿建筑的论文。多年以后，作为事后的评论，梁思成指出："他们都不了解中国建筑的'文法'；他们对于中国建筑的描述都是一知半解的。在两人之中，喜龙仁较好。他尽管粗心大意，但还是利用了新发现的《营造法式》一书。"

由于梁思成的父亲坚持要他们完成学业后再结婚，梁思成和林徽因的婚礼直到 1928 年 3 月才在渥太华举行，当时梁思成的姐夫任当地中国总领事。他们回国途中，绕道欧洲考察旅行，在小汽车里走马观花地把当年学过的建筑物都浏览了一遍，游踪遍及英、法、西、意、瑞士和德国等地。这是他们第一次一同对建筑进行实地考察，而在后来的年月中，这种考察旅行他们又进行过多次。同年仲夏他们突然获知国内已为梁思成找到了工作，要求他立即到职。而直到此时，他们才得知梁思成的父亲病重，不久，梁父便于 1929 年 1 月过早地去世了。

1928 年 9 月，梁思成应聘筹建并主持沈阳东北大学建筑系。在妻子和另外两位宾大毕业的中国建筑师的协助下，他建立了一套克雷特式的课程和一个建筑事务所。中国的东北地区是一片尚待开发的广袤土地，资源丰富。若不是受到日本军国主义的威胁，在这里进行建筑设计和施工本来是大有可为的。这几位青年建筑师很快就为教学工作、城市规划、建筑设计和施工监督忙得不亦乐乎。然而，1931 年 9 月，仅在梁思成到校三年之后，日本人就通过一次突然袭击攫取了东北三省。这是日本对华侵略的第一阶段。这种侵略此后又延续了十四年，而从1937年起爆发为武装冲突。

这个多事之秋却标志着梁思成在事业上的一次决定性转折。当年6月，他接受了一个新职务，这使他后来把自己精力最旺盛的年华都献给了研究中国建筑史的事业。1929年，一位有钱的退休官员朱启钤由于发现了《营造法式》这部书而在北京建立了一个学会，名叫营造学会（以后改称中国营造学社）。在他的推动下这部书被修复和重印，曾在学术界引起了很大的反响。为了解开书中之谜，他曾罗致一小批老学究来研究它，但是这些人和他自己都不懂建筑。所以，朱启钤便花了几个月的时间动员梁思成参加这个学社并领导其研究活动。

学社的办公室就设在天安门内西侧的一排朝房里。1931年秋天，梁思成在这里又重新开始了他早先对这部宋代建筑手册的研究。这项工作看来前途广阔，但他对其中大多数的技术名词仍然迷惑不解。然而，过去所受实际训练和实践经验使他深信，要想把它们弄清楚，"唯一可靠的知识来源就是建筑物本身，而唯一可求的教师就是那些匠师"。他想出这样一个主意，即拜几位在宫里干了一辈子修缮工作的老木匠为师，从考察他周围的宫殿建筑构造开始研究。这里多数的宫室都建于清代（1644—1912）。公元1734年［清雍正十二年］曾颁布过一部清代建筑规程手册——《工程做法则例》，其中也同样充满了生僻难懂的术语。但是老匠师们谙于口述那些传统的术语。在他们的指点下，梁思成学会了如何识别各种木料和构件，如何看懂那些复杂的构筑方法，以及如何解释则例中的种种规定。经过这种第一手的研究，他写成了自己的第一部著作——《清式营造则例》。这是一部探讨和解释清代建筑做法的书（虽然在他看来，这部则例不能同公元1103年刊行的那本宋代《营造法式》相比）。

梁思成就是通过这样的途径，初步揭开了他所谓"中国建筑的文法"的奥秘。但这时他仍旧读不懂《营造法式》和其中那些11世纪的建筑资料，这对他是一个挑战。然而，根据经验，他深信关键仍在于寻找并考察那个时代的建筑遗例。进行大范围的实地调查已提到日程上来了。

梁氏夫妇的欧洲蜜月之行也是某种实地考察，是为了亲眼看一看那些他们已在宾大从书本上学到过的著名建筑实物。在北京同匠师们一起钻研清代建筑的经验，也是一次类似的取得第一手资料的实地考察，无非是没有外出旅行而已。显然，要想进一步了解中国建筑的文法及其演变，只能靠搜集一批年份可考而尽可能保持原状的早期建筑遗构。

梁氏注重实地调查的看法得到了学社的认可。1931年，他被任命为法式部主任。次年，另一位新来的成员刘敦桢被任命为文献部主任。后者年龄稍长于梁思成，曾在日本学习建筑学，是一位很有才能的学者。在此后的十年中，他们两人和衷共济，共同领导了一批较年轻的同事。当然，两人都是既从事实地调查，也进行文献研究，因为两者本来就密不可分。

尚存的清代以前的建筑到哪里去找呢？相对地说，在大城市里，它们已为数不多。许多早已毁于火灾战乱，其余则受到宗教上或政治上敌人的故意破坏。这样，调查便须深入乡间小城镇和荒山野寺。在作这种考察之前，梁氏夫妇总要先根据地方志，在地图上选定自己的路线。这种地方性的史志总要将本地引以为荣的寺庙、佛塔、名胜古迹加以记载。然而，其年代却未必可靠，还有一些建筑遗构读来似颇有价值，但经长途跋涉来到实地一看，却发现早已面目全非，甚至湮灭无寻了。尽管如此，依靠这些地方志的指引，仍可以在广大地区直至全省范围内进行调查，而不致有重要的遗构被漏掉。当然，也有某些发现是根据人们的传说、口头指引，乃至历来民谣中所称颂的渺茫的古建筑而得到的。在20世纪30年代，中国建筑史还是一个未知的领域，一些空前的发现常使人惊喜。

那个时代，外出调查会遇到严重的困难。旅行若以火车开始，则往往继之以颠簸拥挤的长途汽车，而以两轮硬板骡车告终。宝贵的器材——照相机、三脚架、皮尺、各种随身细软，包括少不了的笔记本，都得带上。只能在古庙或路旁小店中投宿，虱子成堆，茅厕里爬满了蛆虫。村边茶馆中常有美味的小吃，但是那碗筷和生冷食品的卫生情况却十分可疑。在华北的某些地区还要提防土匪对无备的旅客进行突然袭击。

只要林徽因能把两个年幼的孩子安排妥当，梁氏夫妇总是结伴同行，陪同他们的常是梁思成培养的一名年轻同事莫宗江，此外就是一个捐行李和跑腿的仆人。他们去的那种地方电话是很罕见的，地方衙门为了和在别处的上司联系，可能会有一部；小城里就再没有其他线路了。这样，在能够对某个重大发现进行详尽考察之前，往往要花费许多时间去找地方官员、佛教高僧和其他人联系、交涉。

当所有这些障碍终于都被克服，这支小小的队伍便可以着手工作了。他们拉开皮尺，丈量着建筑的大小构件以及周围设置。这些数字和画在笔记本上的草图对于日后绘制平面、立面和断面图是必不可少的，间或还要利用它们来制作主体模型。同时，梁思成除了拍摄全景照片之外，还要背着他的莱卡相机攀上梁架去拍摄那些重要的细部。为了测量和拍照，常要搭起临时脚手架，惊动无数蝙蝠，扬起千年尘埃。寺庙的院里或廊下常常立着石碑，上面记着建造或重修寺庙的经过和年代，多半是由林徽因煞费苦心地抄录下来。所有这些宝贵资料都被记在笔记本上，以便带回北京整理、发表。

1932 年，梁思成的首次实地调查就获得了他的最伟大发现之一——坐落在北京以东 60 英里［约 100 公里］处的独乐寺观音阁，阁中有一尊 55 英尺［约 16 米］高的塑像。这座建于公元 984 年的木构建筑及其中塑像已历时近千年而无恙。

1941 年，梁思成曾在一份没有发表的文稿中简述了 30 年代他们的那些艰苦的考察活动：

　　过去九年，我所在的中国营造学社每年两次派出由研究员率领的实地调查小组，遍访各地以搜寻古建遗构，每次二至三个月不等。其最终目标，是为了编写一部中国建筑史。这一课题，向为学者们所未及，可资利用的文献甚少，只能求诸实例。

　　迄今，我们已踏勘十五省二百余县，考察过的建筑物已逾两千。作为法式部主任，我曾对其中的大多数亲自探访。目前，虽然距我们的目标尚远，但所获资料却具有极重要的意义。

　　对于梁、林两人来说，这种考察活动的一个高潮，是 1937 年 6 月间佛光寺的发现。这座建于公元 857 年的美丽建筑，坐落在晋北深山之中，千余年来完好无损，经过梁氏考察，鉴定为当时中国所见最早的木构建筑，是第一座被发现的唐代原构实例。他在本书中关于这座建筑的叙述，虽然简略，却已表达出他对于这座"头等国宝"的特殊珍爱。

　　20 世纪 30 年代中国营造学社工作的一个值得称道的特色，是迅速而认真地将其在古建调查中的发现，在《中国营造学社汇刊》（季刊）上发表。这些以中文写成的文章对这些建筑都做了详尽的记述，并附以大量图版、照片。《汇刊》还有英文目录，可惜当时所出七卷如今已成珍本。

　　本书是梁思成在第二次世界大战末期，在四川省李庄这个偏远的江村中写成的。1937 年夏，在北京沦陷前夕，梁氏一家和学社的部分成员撤离了北京。经过长途跋涉，来到当时尚在中国政府控制下的西南山区省份避难。此时，学社的研究经费已非常困难。[1] 但他们仍在极艰苦的条件下对四川和云南的古建筑做了一些调查。然而，在八年抗战中，封锁和恶性通货膨胀，使他们贫病交加。刘敦桢离开了学社，到重庆中央大学任教；年轻人也纷纷各奔前程。但梁思成和他的家人仍留在李庄，追随他们的，只有他忠实的助手莫宗江和其他少数几人。林徽因患肺结核病而卧床不起。就在这种情况之下，1944 年，梁思成在妻子始终如一的帮助之下完成了这本唯一的英文著作，目的在于向国外介绍过去十五年来中国营造学社所获得的研究

---

[1] 原文不确。当时营造学社只是接受了国立中央博物院的补助经费，为其收集、编制古代建筑资料。——孙增蕃校注

成果。

第二次世界大战结束后，从与世隔绝、饱经忧患的状态下解脱出来的梁思成于1946年受到了普林斯顿和耶鲁两所大学的热情邀请，赴美就中国建筑讲学。他的中文著作早已为西方所知，此时他已成为一位国际知名的学者。他把全家安顿在北京，于1946年春季作为耶鲁大学的访问教授到了美国。这是他一生中第二次，也是最后一次访问美国。

到此为止，我的叙述没有涉及个人关系，这是为了说明梁思成、林徽因作为学科带头人的重要作用而避免干扰。但是，以上所写的大部分内容，都是我作为梁氏的亲密朋友而了解到的第一手材料，我们的友情对于我的叙述有密切关系。1932年夏，我和我的丈夫［费正清］作为一对新婚的学生住在北京，经朋友介绍，我们结识了梁氏夫妇。他们的年龄稍长，但离留学美国的时代尚不远。可能正因为这样，彼此一见如故。我们既是邻居，又是朋友，全都喜好中国艺术和历史。不管由于什么原因，在我和丈夫住在北京的四年期间，我们成了至交。当我们初次相识时，梁思成刚刚完成了他的第一次实地调查。两年后，当我们在暑假中在山西（汾阳县峪道河）租了一栋古老的磨坊度假时，梁氏访问了我们并邀我们做伴，在一些尚未勘察过的地区做了一次长途实地调查。那次的经历使我终生难忘。我们共同体验了那些原始的旅行条件，也一同体验了按照地方志的记载，满怀希望地去探访某些建筑后那种兴奋或失望之情，还有那些饶有兴味的测量工作。我们回到美国之后，继续和他们夫妇书信往还。第二次世界大战期间，我们作为政府官员回到中国任职，同他们的友谊也进一步加深了。当时我们和他们都住在中国西南内地，而日军则占据着北京和沿海诸省。

我们曾到李庄看望过梁思成一家，亲眼看到了战争所带给他们的那种贫病交加的生活。而就在这种境遇之中，既是护士，又是厨师，还是研究所长的梁思成，正在撰写着一部详尽的中国建筑史，以及这部简明的《图像中国建筑史》。他和助手们为了这些著作，正在根据照片和实测记录绘制约七十大幅经他们研究过的最重要的建筑物的平面、立面及断面图。本书所复制的这些图版无疑是梁思成为了使我们能够理解中国建筑史而做出的十分重要的贡献。

当梁思成等迁往西南避难时，他们曾将实地调查时用莱卡相机拍摄的底片存入天津一家银行［的地下金库］以求安全。但是八年抗战结束后他才发现，这无数底片已全部毁于［1939年的］天津大水。现在，只剩下了他曾随身带走的照片。

1947年，当梁思成来到耶鲁大学时，带来了这些照片，还有那些精彩的图纸和这部书的文稿，希望能在美国予以出版。当时，他在耶鲁执教，还在普林斯顿大学讲学并接受了一个名誉学位，此外，他还同一小批国际知名的建筑师一道，担任了联合国总部大厦设计工作的顾问，工作十分繁忙。他曾利用工作的空隙，和我一道修改他的文稿。1947年6月，他突然获知林徽因需要做一次大手术，便立即动身回北京。行前，他把这批图纸和照片交给了我，却带走了那仅有的一份文稿，以便"在回国的长途旅行中把它改定"，然后寄来给我。但从此却音信杳然。

妻子病情的恶化使梁思成忧心忡忡，无心顾及其他。不久，家庭的忧患又被淹没在革命和中国人的生活所发生的翻天覆地的变化之中了。1950年，新生的中华人民共和国请梁思成在国家重建、城市规划和其他建筑事务方面提供意见并参与领导。甚至重病之中的林徽因也应政府之请参与了设计工作，直到1955年［原文误为1954年］她过早地去世为止。

也许正是她的死使得梁思成重新想到了他这搁置已久的计划。他要求我把这些图纸和照片送还给他。我按照他所给的地址将邮包寄给了一个在英国的学生以便转交给他。1957年4月，这个学生来信说邮包已经收妥。但直到1978年秋我才发现，这些资料竟然始终未曾回到梁思成之手。而他在清华大学执教多年后已于1972年去世，却没有机会出版这部附有插图和照片的研究成果。

现在的这本书是这个故事的一个可喜的结局。1980年，那个装有图片的邮包奇迹般地失而复得。一位伦敦的英国朋友为我追查到了那个学生的下落，得到了此人在新加坡的地址。这个邮包仍然原封未动地放在此人的书架上。经过一番交涉，邮包被送回北京，得以同清华大学建筑系所保存的梁思成的文稿重新合璧。虽然这部《图像中国建筑史》的出版被耽搁了三十多年，它仍使西方的学者、学生和广大读者有机会了解在这个领域中国这位杰出的先驱者的那些发现以及他的见解。

于马萨诸塞州坎布里奇

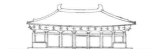

附录 《图像中国建筑史》英文版全文

A PICTORIAL HISTORY Of CHINESE ARCHITECTURE

A Study of the Development of Its Structural System
and the Evolution of Its Types

Liang Ssu-ch'eng
Edited by Wilma Fairbank

Contents

Foreword · · · 216

Acknowledgments · · · 218

Liang Ssu-ch'eng: A Profile · · · 221

Editorial Method · · · 233

Preface · · · 236

The Chinese Structural System

Origins · · · 240

Two Grammar Books · · · 243

Pre-Buddhist and Cave-Temple Evidence of Timber-Frame Architecture

Indirect Material Evidence · · · 253

Han Evidence · · · 254

Cave-Temple Evidence · · · 255

Monumental Timber-Frame Buildings

The Period of Vigor (ca. 850–1050) · · · 259

The Period of Elegance (ca. 1000–1400) · · · 267

The Period of Rigidity (ca. 1400–1912) · · · 274

Buddhist Pagodas

The Period of Simplicity (ca. 500–900) · · · 284

The Period of Elaboration (ca. 1000–1300) · · · 289

The Period of Variety (ca.1280–1912) · · · 295

Other Masonry Structures

Tombs · · · 303

Vaulted Buildings · · · 305

Bridges · · · 306

Terraces · · · 307

P'ai-lou Gateways · · · 308

Chinese Dynasties and Periods Cited in the Text
中国朝代和各时期与公元年代对照表 · · · 310

Glossary of Technical Terms
技术术语一览 · · · 312

Guide to Pronunciation
汉字拼音法指南 · · · 318

Selected Bibliography
部分参考书目 · · · 320

Foreword

Liang Ssu-ch'eng, a leading Chinese architect, was one of the founders of historical research on China's ancient architecture. He wrote this book during the Second World War shortly after completing his fifield investigations in North China and the interior. Professor Liang originally planned it to be one part of a large book, *The History of Chinese Art*. The other part, for which he had already written an outline, was to be on Chinese sculpture, but this plan was never carried out.

As it stands, the book is a valuable short summary of his early work and provides a brief visual survey of the great treasury of ancient Chinese architecture. One can study the "organic" structural system, the evolution of building types, and the development of various architectural elements by the comparative method. For the beginner it is a good introduction to the study of Chinese architectural history. The specialist, too, will draw inspiration from it. Liang never ceased searching for new understanding and is able to explain the profound in simple terms. Above all, the beautiful drawings from the skilled hands of Professor Liang and Mo Tsung-chiang provide a delightful esthetic experience for the reader.

Liang dedicated his life to architecture, and his contributions were many sided. Not only did he leave us many academic treatises and books in Chinese, which are going to be published or reprinted in Peking within the next few years, but he was also a respected and influential educator. He founded two architectural departments—one at

Northeastern University, Liaoning Province, in 1928, and one at Tsing Hua University in 1946, which still flourishes. His students have spread all over China and work in a wide range of fields. From 1949 Liang participated in China's socialist reconstruction with heart and soul. He was appointed as one of the honored leaders in charge of designing the National Emblem of the People's Republic of China and, later, of the Monument of the People's Heroes. In addition, he did much useful work in planning the city of Peking and promoting historical conservation throughout China. Though Liang died more than ten years ago, he is still remembered with great respect and affection.

That this work will finally be published in the Western world as the author wished is especially due to the efforts of Liang's old friend Wilma Fairbank. After his death she helped us to find and recover the precious drawings that had been lost for more than twenty years and carefully shepherded the manuscript and many illustrations to their final form in this book.

Wu Liang-yong
Professor and Dean of the Architectural Department, Tsing Hua University
Member of the Technical Science Division, Chinese Academy of Sciences, Peking

Acknowledgments

Many admirers of Liang Ssu-ch'eng and of Chinese architecture have joined in the efforts to get his long-lost book into the hands of the Western readers for whom he intended it. Professor Wu Liang-yong, head of the Department of Architecture at Tsing Hua University, led the way when he asked me in 1980 to edit the manuscript and find an American publisher for it. I was happy to resume the responsibilities Liang had given me thirty-three years before.

When The MIT Press, long noted for the quality of its architectural books, accepted Liang's book for publication, the project was well started. However, the problems of editing such a complex volume across the Pacific loomed large. Unquestionably our basic good fortune was the whole-hearted cooperation of Liang's second wife, Lin Chu. She is a member of the Department of Architecture at Tsing Hua University, very cognizant of her husband's work, and devoted to his memory. We met in Peking in 1979 and worked together there in 1980 and 1982. In addition to performing her fulltime job, for three years she has shared with me endless details of checking, numbering, and labeling illustrations, supplied me with missing items, and answered my constant questions. Our airmail correspondence, she writing in Chinese, I in English, has been prompt and unceasing. She, a dear friend, deserves my primary gratitude.

After the text and illustrations were reunited in Peking in the summer of 1980, I spent time there on two occasions. The miraculous recovery of the lost materials opened many doors for me. My old friends, Liang's sister Liang Ssu-chuang, his son Liang Congjie and family, and his friend Chin Yueh-lin, were warmly welcoming and extremely helpful. I had the pleasure of interviewing and later corresponding with three elderly architects who had been Liang's classmates at Penn and his lifelong friends, Yang Ting-bao and Tung Chuin, both now deceased, and Chen Chih. Of the younger architectural historians who had participated in the 1930s field work of the Institute, I met Mo Tsung-chiang, Chen Ming-da, Lo Che-wen, Wang Shih-hsiang, and Liu Hsu-chieh, son and follower of Liu Tun-tseng. All had spent the war years in the southwestern provinces of Yunnan and Szechuan where the Institute took refuge. Some of the next younger generation, Liang's postwar students at Tsing Hua from the 1950s onward, have been particularly helpful during the final months of preparing the book for publication, especially Xi Shu-xiang, Yin Ye-ho, and Fu Xi-nian, as well as Fu's associate Sun Chen-fan in the China Building Technology Development Center in Peking. The glossary owes much to all four; Fu and colleagues sent some new photographs; and Xi made explanatory drawings for the editor's note and provided invaluable support in many other ways.

Sir Anthony Lambert and Tim Rock in London played important parts in recovering the lost illustrations. An important influence on me was Else Glahn of Aarhus University, Denmark, who is Europe's leading specialist in Chinese architecture. Sharing my admiration for Liang's work, she cooperated with me in pursuing it and I learned much from her long before beginning this book.

In America I have had generous help from the archivists of the University of Pennsylvania, Princeton, Yale, and Harvard. At Princeton, Robert Thorp and Huang Yun-sheng, a former student of Liang, have lent guidance and moral support. Friends at Yale have been unfailing: Jonathan Spence, Mary's Wright, Mary Gardner Neill, the architect King-lui Wu, and especially Helen Chillman to whom he directed me. She has charge of the unique collection of Chinese architecture slides made from Liang's prints for his 1947 lectures at Yale. Harvard being my home base, I have used the Harvard-Yenching Library and the Fogg Art Museum constantly. Eugene Wu, director of the former, and John Rosenfield, acting director of the latter, deserve my special thanks. William Coaldrake, historian of Japanese architecture, has managed to be always there when I needed him. Architects in the area have been very kind. I think especially of Paul Sun and David Handlin, who were advocates of the project from the first. Robin Bledsoe edited my editing, and my friend Joan Hill typed and retyped the manuscript. My sister Helen Cannon Bond gave me loving ncouragement and practical assistance.

Foundation support from the American Philosophical Society and the National Endowment for the Humanities supported my research and travel. My trips to Peking were not only fruitful but joyous largely due to the hospitality of Arthur and Sheila Menzies and John and Michele Higginbotham of the Canadian Embassy.

My husband, John, stayed at home but that was his only vacation from the world of *tou-kung* that impinged on our lives as the book took shape. As always I thank him for his quiet confidence in me and the expert assistance he provided whenever I needed it.

Wilma Fairbank

Liang Ssu-ch'eng: A Profile

To us in the West the traditional architecture of China has long had the fascination of the exotic. Pagodas and temple buildings with their uptilted roofs, and the window latticework, garden moongates, and camelback bridges of palaces and residences so enchanted European designers of the early eighteenth century as to inspire the invention of "Chinoiserie". Chinese architectural motifs were depicted in wallpaper, painted on porcelain, imitated in furniture, and reproduced recognizably as outdoor features on the estates of the wealthy. This upper-class vogue perhaps reached its height in England in 1763 with the erection of the Kew Gardens pagoda and has never quite died away.

In China these architectural features had been developed by craftsmen over the centuries in response to basic needs, from simply providing shelter at one extreme to instilling religious piety or displaying imperial grandeur at the other. Slighted, for reasons still obscure, as the province of carpenters and masons, architecture did not evoke the intellectual curiosity that leads to scholarly analysis. Not until the twentieth century did the Chinese undertake the study of their own architectural history, and in this the pioneer was Liang Ssu-ch'eng (pronounced Leong Sss-chung).

By inheritance and training Liang (1901–1972) was superbly fitted to the role fate assigned him — leader of the first generation of Chinese architectural historians. He was the eldest son of the famous scholar-reformer Liang Ch'i-ch'ao, devoted to his father and highly impressionable. Knowledge of China's great traditions and its forward movements was deeply inculcated in the son. Liang was slight of stature and by nature observant, curious, careful, and aesthetically sensitive. He loved to draw and was talented musically. Though he had been born in Tokyo where his father was a temporary political refugee, he grew up in Peking. There he received his initial preparation in the two educational spheres that would prove vital for his future achievements. These were, first, the traditional training, supervised by his father, in classical Chinese that he would need for studying old books and deciphering stelae inscriptions, and, second, a firm grounding in English language, Western sciences, and humanities at Tsing Hua College, specifically designed to prepare students for further study abroad. He and his schoolmates were of that remarkable generation of Chinese intellectuals, bilingual and bicultural, who achieved the miracle of bridging the gap between East and West.

Chance played its part in Liang's choice of architecture as a career. It was suggested by the girl who was later to become his wife, Lin Whei-yin, daughter of Lin Ch'ang-min, a scholar-diplomat and well-known poet. When Lin Ch'ang-min was sent to London on an assignment in 1921, he took his daughter along for company. She was barely seventeen [1], very bright, quick, and pretty, and already beginning to evidence the magnetism that was to draw others to her irresistibly throughout the years ahead. She had her father's gift for writing poetry but was equally attracted to the other arts, particularly drama, drawing, and painting. Enrolled in a British girls' school, she quickly developed her knowledge of English to unusual facility in both speaking and writing. From a British schoolmate who played at designing houses, she learned of the profession of architect. Attracted to a lifework that combined daily artistic creativity with immediate usefulness, she decided this was the profession she wanted. On her return to China she easily won Liang to the same decision.

Both Liang and Lin Whei-yin decided to study at the University of Pennsylvania Department of Architecture headed by the eminent Paul Cret from the Ecole des Beaux Arts in Paris. Liang's entry was delayed until the fall term of 1924 by a disastrous motorcycle accident in Peking in May 1923 that caused multiple fractures of his left leg. It never healed properly and remained slightly shorter than his right. As a young man,

---

[1] 疑为 When Lin Ch'ang-min was sent to London on an assignment in 1920, he took his daughter along for company. She was barely sixteen······ ——译校注

Liang was strong and active despite this handicap, but in later years it affected his spine and he sometimes suffered intense pain.

The curriculum at Penn, based on the Beaux-Arts tradition, was designed to produce practicing architects but was equally well suited to prepare architectural historians. Students were required to study the classical orders of Greece and Rome and the monuments of medieval and Renaissance Europe. Their skills were tested by such challenging assignments as drafting restorations of ancient ruins or completing the designs of unfinished cathedrals. A basic requirement was the development of clear and beautifully executed architectural rendering, including the necessary lettering. Liang excelled in this and later required the same high level of performance from his young coworkers and students.

During his second year at Penn, Liang received from his father in Peking a book that was destined to affect his entire life. It was a key to Sung dynasty architecture, the *Ying-tsao fa-shih* (Building Standards) , compiled in 1103 by a brilliant official at the Sung court, which set forth in unfamiliar terminology the techniques of Sung construction. It had essentially disappeared in succeeding centuries but a manuscript copy had recently been discovered and published. Liang tackled the book at once but, he admitted later, understood very little. Beforehand he had apparently given little thought to Chinese architectural history, but from now on the challenge to solve the obscurities of this key work wasfixed in the back of his mind.

Lin Whei-yin reached Penn also for the fall term of 1924, only to discover that women were not admitted to the Department of Architecture. She enrolled in the university's School of Fine Arts, managed to take courses in architecture and in fact, in 1926 was appointed "Part-Time Assistant to the Architectural Design Staff." The following year she was elevated to "Part-Time Instructor." In June 1927 she received the Bachelor of Fine Arts degree with honors at the same ceremony in which Liang was awarded with similar distinction the degree of Master of Architecture.

After a summer spent working in Cret's office in Philadelphia, the pair parted for a few months to pursue separate studies. Lin Whei-yin's long interest in drama led her to

Yale to study stage design at George P. Baker's famous workshop. Liang went to Harvard to study the writings of Western scholars on Chinese art and, if possible, architecture.

At just this period, when Liang was in his mid-twenties, the first books seriously devoted to Chinese architecture appeared in the West. A German, Ernst Boerschmann, published two volumes of photographs of characteristic Chinese building types in 1923 and 1925. Two treatises by the Swedish art historian Oswald Siren on the city walls, gates, and palaces of Peking appeared in 1924 and 1926. Much later, from hindsight, Liang commented, "Neither knew the grammar of Chinese architecture; they wrote uncomprehending descriptions of Chinese buildings. But, of the two, Siren was the better. He used the newly recovered *Ying-tsao fa-shih*, though carelessly."

The wedding of Liang and Lin Whei-yin, deferred at his father's urging until both had completed their studies, took place in March 1928 in Ottawa, where Liang's elder sister was living as wife of the Chinese Consul-General. The young couple left at once for a grand tour of Europe on their way home. Rushing by car from place to place to see in a limited time everything they had studied, they "covered" England, France, Spain, Italy, Switzerland, and Germany. It was the first of the many architectural field trips they were to share in the years to come. In midsummer they were suddenly called home by word of an important job opening for Liang. Only then did they learn of the serious illness of Liang's father, who was shortly to suffer an untimely death in January 1929.

Young Liang was appointed in September 1928 to establish and head a Department of Architecture at Northeastern University in Shenyang, then known to Westerners as Mukden, Manchuria. With the help of his wife and two other Chinese architects from Penn, he set up a Cret-style curriculum and an architectural design firm. China's vast northeast was a still underdeveloped territory of great potential wealth. The demand for architectural design and construction appeared limitless, save for the menace of Japanese militarism. The young architects were soon busy with teaching, city planning, designing, and supervising construction. But in September 1931, three years after Liang had joined the university faculty, a Japanese military coup seized control of the northeastern

图像中国建筑史

provinces. It was the completion of the first stage of Japanese aggression in China, which was to continue for the next fourteen years, exploding into armed conflict in 1937.

This autumn of blighted hopes marked a decisive turn in Liang's career. He had accepted in June a new position, which led to his spending his most productive years as an architectural historian. The Society (later institute) for Research in Chinese Architecture had been established in Peking in 1929 by a wealthy retired official, Chu Ch'i-ch'ien, as a consequence of his discovery of the *Ying-tsao fa-shih*. Its recovery and reprinting at his initiative had caused a sensation in scholarly circles. Hoping to unlock its mysteries, he had assembled a small bookish staff but, since neither they nor he had training in architecture, Chu had for some months been urging Liang to join the Society to direct its research.

The Society's headquarters were rooms just inside the Tien An Men on the west side of the courtyard. There in the autumn of 1931 Liang resumed his earlier study of the Sung manual. It seemed so promising, yet most of the technical terms still eluded him. His practical training and experience told him, as he expressed it, that the only reliable sources of information are the buildings themselves and "the only available teachers are the craftsmen." He conceived the idea of starting his research by studying the construction of the palace buildings around him under the tutelage of certain old carpenters who had spent their lives maintaining them. Most of the buildings had been erected in the Ch'ing (Manchu) dynasty (1644–1912) . A Ch'ing handbook of structural regulations, the *Kung-ch'eng tso-fa tse-li*, had been published in 1734. It was filled with equally unfamiliar technical terms, but the old craftsmen knew orally the traditional terms, and with their guidance Liang was able to identify the various timbers and other structural parts, to observe the complex building methods, and to decipher the regulations cited in the manual. This firsthand research resulted in his first book, *Ch'ing t'ai ying-tsao tse-li* (*Ch'ing Structural Regulations*) , which discussed and explained the Ch'ing manual (although in Liang's view it was not in the same class as the 1103 Sung manual) .

In this way Liang achieved his first insights into what he called "the grammar of Chinese architecture." He was still puzzled and challenged by the *Ying-tsao fa-shih*, with its data about eleventh-century buildings. But experience had convinced him that the key was to find and examine surviving buildings of that period. The time had come for wide-ranging field trips.

The Liang's European honeymoon had been a field trip of a sort, to see firsthand the architectural monuments already studied from secondary sources at Penn. His months of scrutiny of Ch'ing architecture with the carpenters in Peking had resembled the firsthand inspection of field trips without the travel. It was obvious that further comprehension of the grammar of Chinese architecture and of its evolution must depend upon collecting a corpus of extant dated structures of earlier periods that had remained as far as possible in their original condition.

Liang's special interest in field investigations was recognized by the Society, which named him Director of Technical Studies, in 1931. A coordinate position, Director of Documentary Studies, was given the next year to a newcomer, Liu Tun-tseng, a very able scholar a few years older than Liang who had received his architectural training in Japan. The two men worked harmoniously together as a mutually supportive team leading their younger colleagues through the following decade. Inevitably, each did both documentary research and field studies because the two aspects to the work were fundamentally inseparable.

Where were the surviving buildings of pre-Ch'ing date to be found? Relatively few remained in the big cities. Many had been destroyed carelessly by fire or conflict, others intentionally by dynastic or religious enemies. The search had to explore the smaller towns in the countryside and remote temple complexes in the hills. Before undertaking such expeditions, the Liangs learned to map out their routes on the basis of local gazetteers. These scholarly compilations of local history noted the ancient temples, pagodas, and other monuments in which each district took pride. But the dates were not necessarily reliable, and often buildings that sounded like treasures and inspired long side trips were discovered to have been drastically altered or demolished. Nevertheless, these guidebooks provided the means of surveying wide areas, eventually whole provinces, without missing whatever important structures had survived. Of course, certain individual discoveries resulted from pursuing rumors, word-of-mouth suggestions, even obscure monuments praised in traditional folksongs. In the 1930s the field of Chinese architectural history was truly open to the excitement of unprecedented finds.

The same period put serious difficulties in the way of the investigators. Travel, if begun by train, was usually followed by crowded bus over rutted roads, ending often by mule-drawn, two-wheeled springless Peking cart. Precious gear had to be transported—cameras, tripods, measuring tapes, and other paraphernalia, including the indispensable note-books. Lodging had to be found in temples or in local inns, where commonly disease-carrying lice thrived and outhouses teemed with maggots. Roadside teahouses often provided delicious simple fare, but the cleanliness of chopsticks, bowls, and any cold or uncooked food was dubious. In certain areas of North China there was also the risk of attack by bandits who preyed on unsuspecting travelers.

The Liangs traveled together whenever Lin Whei-yin could safely leave their two small children. A younger colleague, often Mo Tsung-chiang who had been trained by Liang, usually accompanied them, as well as a man servant to handle baggage and local errands. Telephones were few. The local *yamen* or headquarters might have one as a link with superiors elsewhere, but there were rarely lines within a town itself. Thus much time was consumed in finding and negotiating with local officials, Buddhist dignitaries, and others before an exciting discovery could be examined.

When at last all these obstacles were overcome, the small team set to work. They used metric tapes to measure both large and small components of the structure, as well as its immediate setting. These measurements, and notebook sketches, would be necessary for later making ground-plan, elevation, and section drawings, and in very special cases for making three-dimensional models. Meanwhile Liang himself was climbing over the structure with his Leica camera, taking photographs of significant details to add to his previously snapped general views. Both measuring and photographing often required erecting temporary scaffoldings, disturbing bats and the dust of ages. The inscriptions on the stone stelae standing in the temple courtyards or porches, which recorded the dates and circumstances of the erection or repair of the building, had to be painstakingly copied, often by Lin Whei-yin. All this precious information was set down in notebooks and carried back to Peking for processing and publication.

Liang's first field trip made in 1932, resulted in one of his greatest discoveries. It was the towering Buddhist structure, Tu-le Ssu, located sixty miles east of Peking, which housed a clay figure fifty-five feet high. Both the wood-frame building, erected in 984, and the statue had survived almost a thousand years in good condition.

In 1941 Liang summarized in unpublished notes the strenuous field explorations of the thirties: For the last nine years, the Institute for Research in Chinese Architecture of which I am a member has been dispatching twice every year, on trips of two or three months' duration, small teams of field workers headed by a research fellow to comb he country for ancient monuments. The ultimate aim is the compilation of a history of Chinese architecture, a subject that has been virtually untouched by scholars in the past. We could find little or no material in books; we have had to hunt for actual specimens.

We have, up till today, covered more than 200 *hsien* or counties in fifteen provinces and have studied more than two thousand monuments. As head of the Section of Technical Studies, I was able to visit most of these places personally. We are very far from our goal yet, but we have found materials of great significance.

For the Liangs, the high point of their years of searching was the discovery in June 1937 of a Buddhist temple, Fo-kuang Ssu, built in 857 A. D. This beautiful structure, located deep in the mountains of northern Shansi province, had endured in good condition for over a thousand years and was recognized by Liang to be the oldest wooden building yet discovered in China and the first one of T'ang date. His description of it on pages 43–44, though brief, betrays his special love for this "treasure of the first order."

A remarkable aspect of the work of the Society in the 1930s was the speed and thoroughness with which the architectural discoveries were published in the Society's quarterly bulletin, *Chung-kuo ying-taso hsueh-she hui-k'an*. The articles, in Chinese, described the buildings in detail and illustrated them with numerous photographs and drawings. Tables of contents were listed in English. Unfortunately, copies of the seven volumes of the bulletin are rare today.

This book was written by Liang during the final years of World War Ⅱ in the remote and backward Yangtze river town of Li-chuang, Szechuan. He and his family, with some members of the Society, had left Peking in the summer of 1937 as the city was falling to the Japanese. They made their way in the great refugee migration to the mountain-ringed southwestern provinces, which were still controlled by the Chinese government. There the Society was made an Institute of the National Academy, Academia Sinica[1], with Liang in charge. Surveys of the architecture of Yunnan and Szechuan were undertaken. But as the eight years of war, isolation, and soaring inflation continued, illness and poverty worsened. Liu Tun-tseng left to join National Central University in Chungking, and younger members of the group scattered in their own interest. Liang remained in Li-chuang with his family, his faithful assistant, Mo Tsung-chiang, and a few others. Lin Whei-yin was confined to bed with tuberculosis. It was in such circumstances that Liang, in 1946, with his wife's help as always, composed this, his only book written in English, to acquaint the outside world with the achievements of the Institute for Research in Chinese Architecture during the previous fifteen years.

Emerging from isolation and privation at the end of the war, Liang received in 1946 cordial invitations from both Princeton and Yale universities to revisit the United States and lecture on Chinese architecture. His Chinese publications had reached the West, and he was now a scholar of international reputation. He settled his family in Peking and accepted the Yale appointment as visiting professor for the spring term of 1947. It was to be his second and last visit to the United States.

Thus far I have written this account impersonally, since its purpose is to present Liang and Lin Whei-yin in their important leadership roles without intrusion. But much

---

[1] 疑为 There the Society began to accept subsidy from the National Central Museum。——译校注

of what I have written above I know firsthand from a close friendship with the Liangs, which has a bearing on this story. My husband and I were young newlywed students in Peking in the summer of 1932 when mutual Chinese friends introduced us to the Liangs. They were several years older, but their student years in America were not very far in the past. Perhaps that prompted our immediate liking for each other. They were neighbors as well as friends, and we all shared interests in Chinese art and history. For whatever reasons, we became intimates in the next four years of our stay in Peking. Liang, at the time we met, had just completed his first field trip. Two years later they visited us in an old mill in Shansi, which we had rented for the summer, and we had the unforgettable experience of accompanying them on a long field trip into hitherto unexplored territory. We shared with them the primitive travel conditions, the excitement or disappointment at the end of promising side trips to look for buildings described in the old gazetteers, and the interesting tasks of making measurements. Our friendship continued by correspondence after our return to America, and it deepened when we returned to China in government service during World War II . At that time we, as well as they, were in the far west of China while the Japanese armies occupied Peking and the entire east coast.

A visit to the Liangs in Li-chuang was a revelation to us of the poverty and misery induced by the war. Yet under these conditions Liang, who was nurse and cook as well as research director, was writing a detailed history of Chinese architecture in Chinese as well as the shorter *Pictorial History*. To illustrate these works he and his draftsmen were preparing from photograph, and field-trip measurements some seventy large drawings— plans, elevations, and sections—of the most important architectural monuments they had studied. The drawings, reproduced here, are unquestionably fundamental contributions by Liang to our understanding of the history of Chinese architecture.

When the eight years of war ended, Liang discovered that the negatives of the Leica photographs that he and his coworkers had taken on their field trips had been destroyed by floods in the Tientsin bank where they had been left for safekeeping when the group went as refugees to the west. Only the prints that he had taken with him remained.

When Liang came to Yale in 1947, he brought the photographic prints, the superb drawings, and the text of his *Pictorial History* in the expectation of finding a publisher. He was kept busy with teaching at Yale, lecturing and receiving an honorary degree at Princeton, and consulting on the design of the UN headquarters with a small group of internationally known architects. In moments snatched here and there he and I worked on his text. In June 1947 he suddenly received word that Lin Whei-yin must have a serious operation, and he departed for Peking at once. Leaving the drawings and photographs with me, he took the sole copy of the text with him "to finish on the long trip home" and mail back to me. It never came.

Alarm and grief at the crisis in his wife's precarious health were enough to distract Liang. The family's personal concerns were soon submerged by the tumultuous events of the revolution and the remaking of Chinese life. In 1950 the newly established People's Republic of China turned to Liang for counsel and leadership in rebuilding, city planning, and other architectural matters. Even the gravely ill Lin Whei-yin participated in designing at the request of the government before her early death in 1954.

Perhaps it was the shock of her death that turned Liang's thoughts back to his long-abandoned project. He sent for the package of illustrations he had left behind. I mailed them at his direction to the address of a student in England for forwarding to him. A letter from the addressee confirmed that they reached England safely and in good condition in April 1957. Not until the autumn of 1978 did I discover that they had never reached Liang, who had died in 1972 after years of influential teaching at Tsing Hua University but without the opportunity to publish his research findings illustrated by the relevant drawings and photographs.

This book is the happy ending of the story. In 1980 the lost package of illustrations was miraculously recovered. A British friend in London managed to trace for me the student's current whereabouts and turn up an address in Singapore. There lay the package, untouched, on a shelf. After some negotiation it was sent off to Peking and reunited with Liang's text in the Department of Architecture of Tsing Hua University.

Though its publication has been delayed for over thirty years, Liang's *Pictorial History of Chinese Architecture* brings to Western scholars, students, and general readers the architectural discoveries and insights of China's outstanding pioneer in this field.

Wilma Fairbank
Cambridge, Mass.

Editorial Method

As editor of this book, I have observed certain self-imposed constraints necessitated by the strange history of the manuscript. Fundamental, of course, is the fact that Liang, who wrote it in the 1940s, died more than a decade before its publication. Had he lived he might well have rewritten and expanded certain parts, notably his very brief comments on bridges, tombs, and other constructions that follow his fuller treatment of timber-frame buildings and pagodas. In the circumstances, my responsibility was to be scrupulously faithful to his ideas, and as far as possible to his choice of words, without making extraneous alterations in the typescript that he left. My one interpolation is clearly identified.

When Liang decided to present to Western readers a summary of the discoveries made by the Institute for Research in Chinese Architecture during its 1931–1937 studies and explorations in North China and its 1938–1946 wartime refuge in Yunnan and Szechuan, he settled on a pictorial history—not of all Chinese architecture as the title might imply but of important monuments of North China and of relevant examples from other provinces studied by the Institute. He intended to tell his story through selected photographs of historic structures from the Institute's files, interpreted by means of drawings showing their plans, elevations, and sections based on field measurements. Instead of an accompanying text, identifications and brief comments in Chinese and English were to be lettered on the drawings themselves.

The drawings were completed as planned in 1943. Liang took them from Li-chuang, Szechuan, downriver to Chungking where friends in the American Office of War Information photo lab recorded them for him on two reels of microfilm. He sent me one reel for wartime safekeeping, which I deposited in the Harvard-Yenching Library. When he brought the original drawings to America in 1947 he left the second reel in China. This precaution was to have interesting results.

Meanwhile, Liang admits "after the completion of the drawings it seemed that a few words of explanation might be necessary." The text, which sets forth his analyses and conclusions based on his years of unprecedented field studies, is brief but deals with a surprisingly large number of structures. The brevity precludes consideration of many aspects of Chinese architecture that Liang treats in his Institute *Bulletin* articles on the same buildings. The headings by period reflect his evaluations at the time of writing. That time was nearly forty years ago, but the book has long been needed, and rather than distort it by updating, it seems best to preserve and present it as a historic document. Two draft pages on imperial gardens at the end have been omitted as incomplete and remote from the theme of structural development.

Liang wrote in clear, simple English and introduced important Chinese architectural terms in Wade-Giles romanization, which is retained, with a few exceptions for proper names varied by personal choice. The glossary of technical terms reminds the reader of their meanings and provides the Chinese characters. This bilingual feature has a prime significance. The captions lettered on the drawings in both languages imply that serious Western students of Chinese architecture must become familiar with technical terms and the names of monuments and localities in Chinese. Liang at Penn learned the analogous English and French terms for Western architectural history, to which he added his own hard-won knowledge of Chinese architectural history. Similarly, future generations of students in the West who will travel easily and often to China will need to know both its language and its hitherto neglected art of architecture.

Most of the photographs show the monuments in their condition in the 1930s.

　　　　　　　　　　　　　　图像中国建筑史

A few buildings reported to be no longer existing are identified in their captions as "Destroyed" whether by deterioration, accident, or intent, we are not sure. In the hand-lettered drawings minor mistakes in English words, romanizations, or dynasty dates are intentionally left unaltered. It is unfortunate that Liang, beset by personal problems, did not end his text with a summation in 1947. But his "evolution" drawings (figs. 20, 21, 32, 37, 38, 63) , which are here published for the first time, clearly summarize key developmental changes that he traced throughout the text. Others of the drawings have been published in the last thirty years in both East and West without attribution to Liang or the Institute. The source is Liang's second 1943 microfilm reel, which was reproduced in 1952 in a *nei-pu* (internal use only) pamphlet for Tsing Hua architecture students. Copies are known to have reached Europe and Britain. No text was attached so perhaps plagiarism cannot be alleged, but the publication of this book must bring the practice to an end.

Wilma Fairbank

Preface

This volume is by no means an all-inclusive history of Chinese architecture. It is an attempt to present, by means of photographs and drawings of many typical specimens, the development of the Chinese structural system and the evolution of its types. My first intention was to do without any text. But after the completion of the drawings, it seemed that a few words of explanation might be necessary.

The Chinese building is a highly "organic" structure. It is an indigenous growth that was conceived and born in the remote prehistoric past, reached its "adolescence" in the Han dynasty (around the beginning of the Christian era) , matured into full glory and vigor in the T'ang dynasty (seventh and eighth centuries) , mellowed with grace and elegance in the Sung dynasty (eleventh and twelfth centuries) , then started to show signs of old age, feebleness, and rigidity, from the beginning of the Ming dynasty (fifteenth century) . Though it is questionable how much longer its lifeblood can be kept flowing, throughout the thirty centuries encompassed in this volume the structure has retained its organic qualities, which are due to the ingenious and articulate construction of the timber skeleton where the size, shape, and position of every member is determined by structural necessity. Thus the study of the Chinese building is primarily a study of its anatomy. For this reason the section drawings are much more important than the elevations. This is an aspect quite different from the study of European architecture, except perhaps the Gothic in which the construction governs more of the exterior appearance than in any other style.

Now, with the coming of reinforced concrete and steel framing, Chinese architecture faces a grave situation. Indeed, there is a basic similarity between the ancient Chinese and the ultramodern. But can they be combined? Can the traditional Chinese structural system find a new expression in these new materials? Possibly. But it must not be the blind imitation of "periods". Something new must come out of it, or Chinese architecture will become extinct.

A study of Chinese architecture would be incomplete without reference to Japanese architecture, for certain early phases of Japanese architecture should properly be classified as Chinese exports. However, this brief text can only touch upon the subject.

The reader should not be surprised that the overwhelming number of architectural examples presented here are Buddhist temples, pagodas, and tombs. In all times and all places religion has provided the strongest impetus to architectural creation.

The materials in this volume are chosen almost entirely from the files of the Institute for Research in Chinese Architecture. Some of them have been published in the bulletin of the Institute. Since its foundation in 1929, under the inspiring guidance of Mr. Chu Ch'i-ch'ien, President, and Dr. Y. T. Tsur, Acting President during the war years, 1937–1946, the Institute has been engaged in systematically searching the country for architectural specimens, studying them from both the archaeological and geographical points of view. So far, more than two hundred *hsien* (districts) in fifteen provinces have been covered. Had it not been for the interruption of the war, during which field work was almost entirely suspended, we would have a much richer collection to present. I am further handicapped by the fact that as I write, in far western Szechuan, many of my materials are not at hand. They were left in Peking when the Institute moved to the hinterland. Meanwhile, a number of the buildings presented here must have been lost forever as a result of the war. The extent of damage can be ascertained only by the reexamination of each individual monument.

The materials of the Institute were collected during numerous field trips, conducted either by Professor Liu Tun-tseng, formerly Director of the Documentary Section of the Institute, and now Chairman of the Department of Architecture and Dean of the College of Engineering, National Central University, or by myself. I am most grateful to Mr. Liu for his permission to use some of his materials in this volume. I am indebted to my colleague, Mr. Mo Tsung-chiang, Associate Research Fellow of the Institute, who accompanied me on almost every one of my field trips, for making most of the plates.

To Dr. Li Chi, Chairman of the Section of Archaeology of the Institute of History and Philology, Academia Sinica, and Mr. Shih Chang-ju, Associate Research Fellow of the same Institute, I am thankful for their permission to reproduce a drawing of the Shang-Yin site at the Anyang excavations; and to Dr. Li again, as Director of the National Central Museum, for his permission to use some of the materials excavated in the Han tombs at Chiang-k'ou, in which the Institute for Research in Chinese Architecture participated.

To my friend and colleague, Wilma Fairbank, member of the Institute for Research in Chinese Architecture, who has traveled extensively in China and participated in one of my field trips, I am indebted not only for the reconstruction of the shrines of the Wu family and of Chu Wei but also for her strong support and encouragement, which greatly accelerated my work on this volume. I am grateful for her patience in reading the text and correcting my faulty English during her very busy days in Chungking as Cultural Relations Officer of the American Embassy, in which capacity she has done invaluable work in furthering the cultural relations between the two countries.

Finally, gratitude is due my wife, colleague, and former classmate, Lin Whei-yin (Phyllis), who for more than twenty years has contributed untiringly to our common cause, slaved for me over our architectural school problems (and I slaved for her, too), accompanied me on most of the field trips, made important discoveries, measured and drew a great number of buildings, and has borne with tact and fortitude, in spite of her ill health in recent years, the major share in keeping up the spirit and morale of the Institute throughout the incredibly hard war years. Without her collaboration and inspiration, neither this volume nor any of my work in the study of Chinese architecture would have been possible.[1]

Liang Ssu-ch'eng
Wartime Station of the
Institute for Research in Chinese Architecture,
Li-chuang, Szechuan
April 1946

[1] Most of the individuals to whom Liang extended his special thanks in 1946 are deceased.
Surviving in addition to myself are Mo Tsung-chiang at Tsing Hua University, Peking, and Shih
Chang-ju at Academia Sinica, Taipei. —Ed.

The Chinese Structural System

Origins

The architecture of China is as old as Chinese civilization. From every source of information—literary, graphic, exemplary—there is strong evidence testifying to the fact that the Chinese have always employed an indigenous system of construction that has retained its principal characteristics from prehistoric times to the present day. Over the vast area from Chinese Turkestan to Japan, from Manchuria to the northern half of French Indochina, the same system of construction is prevalent; and this was the area of Chinese cultural influence. That this system of construction could perpetuate itself for more than four thousand years over such a vast territory and still remain a living architecture, retaining its principal characteristics in spite of repeated foreign invasions—military, intellectual, and spiritual—is a phenomenon comparable only to the continuity of the civilization of which it is an integral part.

Near Anyang, Honan Province, at the site of the palaces and necropolis of the Shang-Yin emperors (ca. 1766-ca. 1122 B. C.) , excavated by the Academia Sinica, are found the earliest known remains of buildings in China (fig. 10) . These are large loess platforms, upon which undressed boulders, flat sides up, are placed at regular intervals, each topped by a bronze disc (later known as a *chih*) . On top of these discs are found charred logs, the lower ends of wooden posts that once supported the superstructures that were burned at the sack of the capital when the Yin dynasty fell to the Chou conquerors (ca. 1122B. C.). The arrangement of these column bases testifies to the existence of a structural system that had by this time taken a definite form and was destined to provide shelter for a great people and their civilization from that time until today.

The basic characteristics of this system, which is still used, consist of a raised platform, forming the base for a structure with a timber post-and-lintel skeleton, which in turn supports a pitched roof with overhanging eaves (figs. 1, 2) . This osseous construction permits complete freedom in walling and fenestration and, by the simple adjustment of the proportion between walls and openings, renders a house practical and comfortable in any climate from that of tropical Indochina to that of subarctic Man-churia. Due to its extreme flexibility and adaptability, this method of construction could be employed wherever Chinese civilization spread and would effectively shelter occupants from the elements, however diverse they might be. Perhaps nothing analogous is found in Western architecture, with the limited exception of the Elizabethan half-timber structure in England, until the invention of reinforced concrete and the steel framing systems of the twentieth century.

Editor's Note: The Curved Roof and Bracket Sets

Figures 1 and 2 present the basic characteristics of traditional Chinese architecture in a way that will be clear to a prepared mind. However, since not every reader will have had the opportunity to see in situ or to study Chinese timber-frame buildings, a few words of further explanation are offered here.

The immediately outstanding feature of Chinese monumental architecture is the curved roof with overhanging eaves, which is supported by a timber skeleton based on a raised platform. Figure 3 illustrates nine variations of the five types of roof construction listed by Liang on page 26. To understand how and why these curved roofs with their upturned eaves are constructed, we must examine the timber skeleton itself. In Liang's words, "the study of the Chinese building is primarily a study of its anatomy. For this reason the section drawings are much more important than the elevations."

The section drawings show us that the roof supports in Chinese timber-frame construction differ fundamentally from the conventional Western triangular roof trusses that dictate the rigidity of our straight pitched roofs. The Chinese frame is instead, markedly flexible (fig. 4) . The timber skeleton consists of posts and cross beams rising toward the ridge in diminishing lengths. The purlins—horizontal members that support the rafters—are positioned along the stepped shoulders of the skeleton. The rafters are short, stretching down only from purlin to purlin. By manipulating the heights and widths of the skeleton, a builder can produce a roof of whatever size and curvature are required. The concave curved roof allows the semicylindrical rooftiles to fit together snugly for watertightness.

The extent of eave projection is also remarkable. For example, the eaves of the T'ang temple Fo-kuang Ssu (857 A. D.) , which Liang discovered, project about fourteen feet out from their supporting columns (fig.24) . Their importance in sheltering the wooden structure from weather damage for over 1100 years is obvious—for instance, they throw away from the building the rainwater that courses down the tile troughs of the concave roof.

But the immediate function of raising the roof edges is to permit light to penetrate to the interior of the building despite the wide overhang. This necessitates both extending support far outside the interior skeleton to carry the overhanging eave and also building up this support vertically to handle the upturn. How is this achieved?

As Liang points out, "*the tou-kung* [bracket set] plays the leading role, a role so important that no study of Chinese architecture is feasible without a thorough understanding of this element, the governing feature of the Chinese 'order'." Figure 5 is an isometric view of a bracket set, and figure 6 shows such a *tou-kung* (pronounced doe-goong) in place atop a supporting column. Again we are encountering an exotic element. We in the West are accustomed to simple capitals that receive a direct weight and transfer it to the column. The *tou-kung* is a very complex member. Though its base is simply a large

　　　　　　　　　　　　图像中国建筑史

square block on the top of the column, there are set into that block crossed arms (*kung*) spreading in four directions. These in turn bear smaller blocks (*tou*) that carry still longer arms spreading in the four directions to support upper members in balance. The jutting arms (*hua-kung*) rise in tiers or "jumps" and extend outward in steps from the large-block fulcrum to support the weight of the overhanging eaves. This external pressure is countered by internal downthrusts at the other ends of the bracket arms. Intersecting the *hua-kung* in the bracket set are transverse *kung* that parallel the wall plane. Long cantilever arms called *ang* descend from the inner superstructure, balance on the fulcrum, and extend through the bracket sets to support the outermost purlins (fifig.6) . This outer burden is countered by the downthrust of upper interior purlins or beams on the "tails" of the *ang*. The extruding "beaks" of the *ang* easily identify them in the bracket sets. Liang explains more fully these and other complexities in the evolution of this construction. — Ed.

Two Grammar Books

As this system matured through the ages, a well-regulated set of rules governing design and execution emerged. To study the history of Chinese architecture without a knowledge of these rules is like studying the history of English literature before learning English grammar. Therefore, a brief examination of them is necessary.

For this purpose we are fortunate to have two important books left to us from two epochs of great building activities: the *Ying-tsao fa-shih* (Building Standards) of the Sung dynasty (960 – 1279) and the *Kung-ch'eug tso-fa tse-li* (Structural Regulations) of the Ch'ing dynasty (1644 – 1912)—two "grammar books" on Chinese architecture, as we may call them. Both government manuals, they are of the greatest importance for the study of the technological aspects of Chinese architecture. We owe to them all the technical terms that we know and all the criteria that we employ today for the comparative study of the architecture of different periods.

The Ying-tsao fa-shih

The *Ying-tsao fa-shih* (Building Standards)was compiled by Li Chieh, superintendent of construction at the court of Emperor Hui-tsung (ruled 1101- 1125) of the Sung dynasty. Of the thirty-four chapters thirteen are devoted to rules governing the design of foundations, fortifications, stone masonry and ornamental carving, "major carpentry" (structural framing, columns, beams, lintels, ties, brackets, purlins, rafters, etc.) , "minor carpentry" (doors, windows, partitions, screens, ceilings, shrines, etc.) , brick and tile masonry (official rank and use of tiles and ornaments) , painted decoration (official rank and design of ornamental painting) . The rest of the text contains definitions of terms and data for estimation of materials and labor. The last four chapters contain drawings illustrating various kinds of designs in carpentry, stonework, and ornamental painting.

The book was published in 1103. During the eight and a half centuries that have since elapsed, both technical terms and forms have changed, and in an atmosphere in which scholars and literary men looked upon technical and manual work with contempt, the book receded into obscurity and was treasured only by connoisseurs as a rare oddity. It is extremely difficult for the layman today to understand, and many of its passages and terms are almost meaningless. Due to the conscientious efforts of the Institute for Research in Chinese Architecture, first by mastery of the regulations of the Ch'ing dynasty (*the Kung-ch'eng tso-fa tse-li*) , and later with the discovery of a considerable number of wooden structures dating from the tenth to the twelfth centuries, many mysteries of the book have been clarified, rendering it now quite "readable."

Because timber is the principal material used in Chinese architecture, the chapters on "major carpentry" are the most important part of the book for understanding the structural system. The essentials of these rules are explained visually in figures 2 and 7 and may be summed up as follows.

1. Modules （*ts'ai and ch'i*）

The term *ts'ai* has a twofold meaning:

（a）A standard-sized timber used for the *kung*, or "arms" of a set of brackets （*tou-kung*）; and all timbers of the same depth and width. There are eight sizes, or grades, of *ts'ai*, which are determined by the type and official rank of the building to be erected.

（b）A module for measurement, described as follows:

The depth of each *ts'ai* is to be divided into 15 equal parts, called *fen*, and the width of the *ts'ai* is to be ten *fen*. The height and breadth of every building, the dimensions of every member in the structure, the rise and curve of the roofline, in short, every measurement in the building, is to be measured in terms of *fen* of the grade of *ts'ai* used.

When two *ts'ai* are used one above another, it is customary to cushion them by filling the gap with a block six *fen* in height, called a *ch'i*. A member measuring one *ts'ai* and one *ch'i* in depth is called a *tsu-ts'ai*, or "full *ts'ai*." The measurements of, and the proportions between the different parts of a building of the Sung dynasty are always expressed in terms of the *ts'ai*, *ch'i*, and *fen* of the grade of *ts'ai* used.

2. Bracket Sets （*tou-kung*）

A set of *tou-kung*, or brackets, is an assemblage of a number of *tou*（blocks）and *kung* （arms）. The function of the set is to transfer the load from the horizontal member above to the vertical member below. A set may be placed either on the column, or on the architrave between two columns, or on the corner column. Accordingly, a set of *tou-kung* may be called a "column set," "intermediate set" or "corner set" depending on the position it occupies. The members that make up a set may be divided into three main categories: *tou*, *kung*, and *ang*. There are four kinds of *tou* and five kinds of *kung*, determined by their functions and positions. But structurally the most important members of a set are the *lu-tou*, or major bearing block, and the *hua-kung*, or arms extending out from it to form cantilevers to both front and rear, at right angles to the facade of the building. Sometimes a slanting member, at approximately a 30-degree angle to the ground, is placed above the *hua-kung*; it is called the *ang*. The "*tail*" or upper end of the *ang* is often held down by the weight of the beam or the purlin, making it a lever arm for the support of the large overhang of the eave.

The *hua-kung* may be used in successive tiers, each extending front and rear a certain distance beyond the tier below. Such a tier and extension is called a *t'iao*, or "jump", and the number of *t'iao* in a set may vary from one to five. Transverse *kung* intersect the *hua-kung* in the *lu-tou*. One or two tiers of transverse *kung* may be used in a *t'iao*. Such an arrangement is known as *chi-hsin*, or "accounted heart," while a *t'iao* without transverse *kung* is known as *t'ou-hsin*, or "stolen heart." One tier of transverse *kung* is called *tan-kung*, or single *kung*; a double tier is called *ch'ung-kung*, or double *kung*. By varying the number of "jumps," by "accounting" or "stealing" the "hearts," by cantilevering with *hua-kung* or with ang, and by using single or double transverse *kung*, different combinations of the *tou-kung* can be assembled.

3. Beams

The size and shape of a beam varies according to its function and position. The beams below a ceiling are called *ming-fu*, or "exposed beams." They are either straight or slightly arched; the latter is called *yüeh-liang*, or "crescent-moon beam." Above the ceiling *ts'ao-fu*, or "rough beams," are used to receive the load of the roof. The circumference of a beam may vary according to its length, but the cross section always retains, as a norm, a ratio of 3 : 2 between its depth and width.

4. Columns

Rules governing the length and the diameter of a column are rather loose. The diameter may vary from one *ts'ai*-plus-one *ch'i* to three *ts'ai*. The column may be either straight or shuttle-shaped. The latter is given an entasis on the upper third of the shaft. The most important rules in columniation are: (1) a gradual increase in the height of the columns from the central bay toward the corners of the building, and (2) a slight in ward incline of the columns, about 1 ∶ 100. These refinements help give an illusion of stability.

5. The Curved Roof（*chu-che*）

The profile of the roof plane is determined by means of a *chu*, or "raising" of the ridge purlin, and a *che*, or "depression" of the rafter line. The pitch is determined by the "raise" of the ridge, which may make a slope varying from 1 ∶ 2 for a small house to 2 ∶ 3 for a large hall, with gradations in between. The height of the raise is called the *chu-kao*. The curve of the rafter line is obtained by "depressing" or lowering the position of the first purlin below the ridge, by one-tenth of the height of the *chu-kao*, off a straight line from the ridge to the cave purlin. Another straight line is then drawn from this newly plotted point to the cave purlin, and the next purlin below is "depressed" by one-twentieth of the *chu-kao*. The process is repeated, and each time the "depression" is reduced by half. The points thus obtained are joined by a series of straight lines and the roof line is plotted. This process is called *che-wu*, or "bending the roof."

Besides these basic rules, Sung methods for the use and shaping of architraves, ties, hip rafters, purlins, common rafters, and other elements are carefully specified in the *Ying-tsao fa-shih*. A careful study of the evolution of the major carpentry of different periods introduced in later chapters gives a fairly clear idea of the development of the Chinese structural system through the ages.

The chapters on minor carpentry give rules for the designing of doors, windows, partitions, screens, and other nonstructural elements. The tradition is generally carried on in later ages without drastic modification. Ceiling coffers are either square or rectangular, and the large principal coffer is often decorated with miniature *tou-kung*. Even Buddhist and Taoist shrines are highly architectural in character and are decorated with *tou-kung*.

The chapter on tile and tile ornaments specifies the sizes and numbers of the ornamental dragon heads and "sitting beasts" that decorate the ridges, according to the official rank and size of the building. The making of a ridge by piling up ordinary roof tiles, though still practiced today in southern China, is a method no longer used in monumental architecture.

The chapters on decorative painting specify the kinds of painting for different ranks of buildings. Rules also govern the distribution of colors, mainly on the principle of contrasting warm and cool colors. We also learn that the shading of colors is obtained by the juxtaposition of colors related in the chromatic scale rather than by deepening a single color. The principal colors used are blue, red, and green, accented with black and white. Yellow is occasionally used. This tradition has been followed from the T'ang dynasty (618–907) to the present.

A great deal of attention is also paid to the details of parts and members, such as the shaping and curving of the *tou*, *kung*, and *ang*, the arching of the beam and the pulvination of its sides, the carving of ornaments on pedestals and balustrades, and the color schemes of various types and grades of decorative painting. In many respects the *Ying-tsao fa-shih* is a textbook in the modern sense of the word.

The Kung-ch'eng tso-fa tse-li

The *Kung-ch'eng tso-fa tse-li* (Structural Regulations) was published in 1734 by the Ministry of Construction of the Ch'ing dynasty. The first twenty-seven chapters are rules for constructing twenty-seven kinds of buildings, such as halls, city gates, residences, barns, and pavilions. The size of each structural member in each building type is carefully specified, differentiating this book from the *Ying-tsao fa-shih*, which gives general rules and ratios for designing and computation. The next thirteen chapters specify the dimensions of each kind of *tou-kung* and the sequence of assembling them. Seven chapters treat doors, windows, partitions, screens, and stone, brick, and earth masonry. The last twenty-four chapters are rules for the estimation of materials and labor.

The only drawings are twenty-seven cross sections of the twenty-seven building types described. There are no instructions for details, such as the shaping of the *kung*

图像中国建筑史

and *ang*, decorative paintings, and the like, which would have been permeated with the characteristics of the time. This drawback, however, is fortunately overcome by the existence of numerous examples of Ch'ing dynasty architecture that can be conveniently studied.

From the first forth-seven chapters of the *Kung-ch'eng tso-fa tse-li* a number of principles can be derived, of which the following, explained visually in figure 8, are the most important ones concerning major carpentry or structural design.

1. Reduction in the Depth of the *Ts'ai*

As noted above, the depth of the *ts'ai* of the Sung style is 15 *fen* (with a width of ten *fen*) and the *ch'i* measures six *fen*, resulting in a *tsu-ts' ai*, or "full *ts'ai*," of 21 *fen* in depth. However, the concept of *ts'ai*, *ch'i*, and *fen* does not seem to have existed in the minds of the builders of the Ch'ing dynasty. The *tou-k'ou*, or "mortise of the *tou*" [literally "block mouth"] , for receiving the *kung* is specified as the Ch'ing module. It is equivalent to the width of the *kung*, and therefore, of the *ts'ai* (ten *fen* in terms of Sung construction) . The dimensions and proportions of the parts of a *tou-kung* bracket are now expressed in multiples or fractions of the *tou-k'ou*. While retaining the six *fen* (now 0.6 *tou-k'ou*) for the gap between the upper and lower *kung*, the depth of the *kung*, or *ts'ai*, itself is reduced from 15 to 14 *fen* (now 1.4 *tou-k'ou*) . The *tsu ts'ai* is then reduced to only 20 *fen*, or two *tou-k'ou* in depth.

There is another major difference between Sung and Ch'ing bracket sets. In the Sung period the major block (*lu-tou*) , in which the cross arms rest on the centerline of columns parallel to the facade, supports several tiers of *ts'ai* with open or plaster-filled gaps between. These tiers are cushioned by *tou* blocks. In the Ch'ing set the tiers of *tsu-ts'ai* measuring one × two *tou-k'ou*, are laid one directly on top of another. Thus the *ch'i*, or gaps allowing space for the *tou*, are eliminated. These seemingly trifling modifications affect the general aspect of the *tou-kung* so much that the difference is immediately apparent by even a casual glance.

2. A Specific Ratio between the Diameter and the Height of the Column

The Ch'ing regulations specify the diameter of a column as six *tou-k'ou* (four *ts'ai* in Sung terms) , and its height as 60 *tou-k'ou* or ten diameters. The diameter of a column of the Sung dynasty, according to the *Ying-tsao fa-shih*, never exceeded three *ts'ai*, and the height was left to the discretion of the designer. Thus proportionally the Ch'ing column is much enlarged and the *tou-kung* drastically reduced in size, dwindling into insignificant pettiness. An unprecedented increase in the number of intermediate bracket sets results. In some cases seven or eight intermediate sets are perched on the architrave between two columns; in the Sung dynasty the number never exceeded two, both according to the *Ying-tsao fa-shih* and as evidenced in existing buildings.

3. The Length and Width of a Building Determined by the Number of *Tou-kung*

As the number of intermediate sets is increased, the distance between them is strictly specified as 11 *tou-k'ou* center to center. Consequently the distances between columns, and therefore the length and width of a building, must necessarily be determined by multiples of 11 *tou-k'ou*.

4. All Facade Columns of Equal Height

The Sung practice of increasing the height of the columns toward the corners of the building was discontinued. The columns, though slightly tapered, are straight with no entasis. Thus a Ch'ing building as a whole presents a more rigid appearance than a Sung structure. The slight in ward incline, however, is still the rule.

5. Increase in the Width of the Beams

Beams of the Sung dynasty generally have a ratio of 3 : 2 between their depth and width. In the Ch'ing rules the ratio is changed to 5 : 4 or 6 : 5, betraying an obvious ignorance of mechanics and of the strength of materials. Moreover, the overall rule of making the beam "two inches [*fen*] wider than the diameter of the column" seems most arbitrary and irrational. All beams are straight; the "crescent-moon beam" has no place in official Ch'ing architecture.

6. Steeper Pitch of the Curved Roof

What is called in the Sung dynasty *chü-che* ("raise-depress") is known in the Ch'ing dynasty as *chü-chia*, or "raising the frame." The two methods, though bringing about more or less similar results, are entirely different in their basic conceptions. The height of the ridge purlin of the Sung structure is predetermined, and the curvature of the roofline is achieved by "depressing" the successive pulins below. The Ch'ing builder starts from the bottom, giving the first *pu* or "step" (that is, the distance between the two lowest purlins) a " *five chü*," or a pitch of 5:10; the second step a "*six chü*," or 6:10; the third, 6½:10; the fourth, 7½:10; and up to "*nine chü*," or 9:10. The position of the ridge purlin becomes the incidental outcome of the successive steps. The pitch of the Ch'ing roof thus obtained is generally steeper than that of the Sung roof, giving a convenient clue for identifying and dating.

As we shall see below, a Ch'ing building is generally characterized by harshness and rigidity in the lines of the posts and lintels, in the excessively steep pitch of the roof, and in the smallness of the *tou-kung* under the eave. Perhaps it was the uncompromising strictness of the dimensions given in the *Kung ch'eng tso-fa tse-li* that succeeded in effacing all the suavity and elegance that we find so charming in a building from the period of the *Ying-tsao fa-shih*.

Neither book mentions ground plans. The *Ying-tsao fa-shih* contains a few plans, but they show columniation, not the internal division of spaces. Unlike European buildings, the Chinese building, whether house or temple, is rarely planned by subdividing the individual unit. Since it may be subdivided so easily by means of wooden partitions or screens between any two columns, the problem of internal planning hardly exists. Planning, instead, concerns the external grouping of individual units. The general principle is the arrangement of buildings around a courtyard, or rather an arrangement that forms a courtyard or patio. The buildings are sometimes connected by colonnades, but these are eliminated in smaller houses. A large house is composed of a series of courtyards along a common axis, and deviation from this principle is rare. The same principle is applied to both religious and secular architecture. In plan there is no basic difference between a temple and a residence. Thus it was not an uncommon practice in ancient times for a high official, or even a rich merchant, to donate his residence to the service of Buddha and have it consecrated as a temple.

Pre-Buddhist and Cave-Temple Evidence of Timber-Frame Architecture

Indirect Material Evidence

The Shang-Yin remains near Anyang mentioned in the previous chapter are merely ruins (fig. 10) . It is by inference that we come to the conclusion that the Chinese structural system was basically the same then as today. For proof we have to turn to examples from later periods.

The earliest source that reveals the appearance of a timber-frame building is the relief on a bronze vase of the period of the Warring States (403–221 B.C.) of the Chou dynasty (fig. 9) . It shows on a platform a two-storied building with columns, eaved roofs, doors, and railings. From a structural point of view, this building must have had a plan essentially the same as the remains at the Shang-Yin site. Particularly significant are the indications of columns with the characteristic bracket sets, or *tou-kung*, which were later formulated into strict proportions very much like the orders in European architecture. Aside from this specimen and the Anyang remains, and perhaps a few other rather insignificant and inadequate representations on bronzes and lacquers, architecture before the Christian era in China is veiled in obscurity. It is highly doubtful that future archaeological excavations will be able to shed much light on the appearance of the superstructure of Chinese buildings in a period of such remote antiquity.

Han Evidence

The earliest remains of real architectural importance are to be found in the tombs of the Eastern Han dynasty (25–220 A.D.) . These monuments may be classified into three categories: (1) rock-cut tombs (fig. 11) , some highly architectural, mostly in Szechuan Province and a small number in Hunan Province; (2) freestanding pylonlike monuments, called *ch'üeh* (fifig.12) , usually in pairs, flanking the entrance to the avenue leading to a palace, temple, or tomb; (3) small shrines in the form of houses (fig.13) , usually placed in front of the tumulus of a tomb. All these relics are of stone, but the essentials of the *tou-kung* and the trabeated construction are so faithfully represented that a fairly clear idea of the wooden structures after which these monuments are modeled can be gathered from them.

Salient characteristics of the architecture of this period, as seen in these three kinds of remains, are: (1) the octagonal column, capped by a very large *tou* or block that often shows a filletlike molding at its bottom, representing the *ming-pan*, a square, small board [1] ; (2) the *kung* arm with an S-curve, which seems unlikely and therefore may not have been the actual way of shaping timbers at that time; (3) roof and eaves supported by rafters and covered by tubular tiles with ridge ornaments.

Of the wooden palaces and houses of the Han dynasty, whose magnificence we can now know only from contemporary odes and essays, there is not a single standing example. But from clay models of houses buried in tombs (fig. 14) and from reliefs decorating the walls of the tomb shrines (fig. 15) , it is not difficult to form an idea of Han domestic architecture. There are both multistoried mansions and modest houses of the common people. There is one example of an L-shaped elevation, with a walled courtyard in the angle of the L. (fig. 16) . We even detect in a watch tower the forerunner of the Buddhist pagoda. The timber-frame system of construction is clearly represented in some of the models. Here again the *tou-kung* plays the leading role, a role so important that no study of Chinese architecture is feasible without a thorough understanding of this

[1] This element disappeared from Chinese architecture after the sixth or seventh century. The term *ming-pan* is borrowed from the Japanese term designating the element in Japanese architecture of the seventh and eighth centuries. —Ed.

element, the governing feature of the Chinese "order." We also find all five kinds of roof construction used in later ages: the hip roof, the flush gable roof, the overhanging gable roof, the gable and hip roof, and the pyramidal roof (fig. 3). The use of the tubular tile was evidently already as common as it is today.

Cave-Temple Evidence

Buddhism reached China at approximately the beginning of the Christian era. Though there are records of the erection of a Buddhist pagoda as early as the beginning of the third century A.D., described as "a multi-storied tower surmounted by a pile of metal discs," we possess today no Buddhist monument before the middle of the fifth century. However, from then on until the later fourteenth century, the history of Chinese architecture is chiefly the history of Buddhist (and a few Taoist) temples and their pagodas.

The Caves of Yun-kang (450–500 A.D.), near Ta-t'ung, Shansi Province, though unquestionably Indian in origin, with prototypes in Karli, Ajanta, and elsewhere in India, show surprisingly little influence from their land of origin. The architectural treatment of the caves is almost exclusively Chinese. The only indications of foreign influence are the idea of the cave temple itself and the Greco-Buddhist motifs in the ornaments, such as the acanthus leaf, the egg-and-dart, the swastika, the garland, the bead, and others, which have enriched and become a part of the vocabulary of Chinese ornamental motifs (fig. 17).

The architecture of the Yun-kang Caves may be studied from two points of view: (1) the caves themselves, including the architectural treatment of the exteriors and interiors, and (2) contemporary wooden and masonry architecture represented in the reliefs decorating the cave walls (fig. 18). Among these depictions are numerous halls and pagodas, structures that once rose in great numbers all over the plains and hills of northern and central China.

The hewing out of cave temples from rocky cliffs was practiced in the north until the middle of the T'ang dynasty (618–907) , and carried on thereafter in the southwest, notably in Szechuan Province, until the Ming dynasty (1368–1644) . Only the earlier cave temples are of interest to the architectural historian. The apogee of their architectural quality was reached in the caves of the Northern Ch'i and Sui dynasties (late sixth and early seventh centuries) at T'ien-lung Shan, near T'ai-yuan, Shansi Province, and at Hsiang-t'ang Shan, on the border of the provinces of Honan and Hopei (fig. 19) .

These caves have preserved in stone faithful copies of the wooden architecture of their time. We notice among the salient characteristics that the columns in most cases are octagonal, with capitals in the shape of a *lu-tou*, similar to those found in the rock-cut tombs of the Han dynasty. Above the capitals is placed the architrave, which, in turn, is to receive the *lu-tou* of the set of *tou-kung*. This arrangement was in later ages modified by mortising the architrave directly onto the upper end of the column, and (combining the two *lu-tou* into one) placing the *lu-tou* of the bracket set directly on the column.

In the treatment of these caves, the *tou-kung* is always the dominant architectural feature. It is still simple, as in the Han dynasty, but the S-curve of the *kung* has by now been straightened into a more plausible form. An intermediate set, in the form of an inverted letter V, is introduced on the architrave spanning the two columns. Because it resembles the Chinese character for man (*jen*) , it is sometimes dubbed *jen-tzu kung*, "man-character arm." With the exception of one surviving specimen, the Tomb Pagoda of Ching-tsang (746) in the Sung Mountains, Honan Province (figs. 64 d, c) , where this feature is represented in brick, it is now preserved in only a few contemporary wooden structures in Japan.

Monumental Timber-Frame Buildings

Timber, which is the principal building material employed by the Chinese, is highly perishable, subject to natural decay from the elements and from pests. It is highly flammable and, when used for a religious building, is constantly exposed to the danger of fire from the incense and candles of the worshippers. Moreover, a land periodically razed by civil wars and religious struggles proved none too favorable for the conservation of timber structures. The customary sack of the capital of the vanquished by the founder of a new dynasty, who was usually either a rebel, a warlord, or a leader of a less developed nation from the north, usually with great animosity toward the defeated ruler, has deprived posterity of every trace of the glory and splendor of every single palace of numberless princes, kings, and emperors. (Among the very few notable exceptions to this barbarous custom was the founding of the Republic in 1912: the palaces of the Ch'ing emperors were opened to the public as a museum.)

Although China has always been considered a land of freedom of worship, there were at least three great persecutions of Buddhism from the fifth to the ninth centuries, The last of these outrages occurred in 845 and almost stripped the country of all Buddhist temples and monasteries. These practices, and the highly perishable nature of the material, probably account for the total absence of timber structures before the middle of the ninth century.

Recent tendencies in China, especially since the founding of the Republic, have not been favorable for the preservation of ancient edifices. After repeated defeats by the modern powers from the middle of the nineteenth century, the Chinese intellectual and governing class lost confidence in everything of its own. Its standard of beauty was totally confused; the old was thrown away; of the new, or Western, it was ignorant. Buddhism and Taoism were condemned as sheer superstition and, not unjustly, as among the causes for the sluggishness of the people. The general tendency was iconoclastic. A great many temples were confiscated and secularized and were utilized by antitraditional officials as schools, offices, grain storages, or even barracks, ammunition dumps, and asylums. At best the buildings were remodeled to suit the new functions, while at worst the unfortunate structures were put at the disposal of ill-disciplined and underpaid soldiers, who, for lack of proper fuel, tore down every removable part-partitions, doors, windows, railings, and even *tou-kung*—for the cooking of their meals.

It was not until late in the twenties that Chinese intellectuals began to realize the significance of their own architecture as an art no less important than calligraphy and painting. First, a number of buildings appeared, built by foreigners but in the Chinese style. Next, Western and Japanese scholars published books and articles on Chinese architecture. And, finally, a number of Chinese students who had gone abroad to study Western techniques of building came back with the realization that architecture is something more than brick and timber: it is an art, an expression of the people and their times, and a cultural heritage. The contempt of the educated class for matters of "masonry and carpentry" gradually turned into appreciation and admiration. But it was a long time before this consciousness could be implanted in the minds of the local authorities upon whom the protection of antiquities depends. Meanwhile the process of deterioration and destruction steadily progressed in the hands of the ignorant and the negligent.

Finally, the destruction brought about by the war of Japanese aggression (1937–1945)is still unknown. It should be no surprise to find that many of the monuments recorded in this volume will hereafter be known only by these photographs and drawings.

Wooden structures existing today, or rather known to have existed when they were last seen in the 1930s, may be tentatively divided into three main periods: the Period of Vigor, the Period of Elegance, and the Period of Rigidity (figs. 20, 21) .

The Period of Vigor includes the phase from the middle of the ninth century to the middle of the eleventh century, or from the reign of Emperor Hsuan-tsung of the T'ang dynasty to the end of the reign of Emperor Jen-tsung of the Sung dynasty. This period is characterized by a robustness of proportion and construction which must have been that of the glorious T'ang dynasty and of which this is but its magnificent conclusion.

The Period of Elegance extended from the middle of the eleventh century to the end of the fourteenth century or from the reign of Emperor Ying-tsung of the Sung dynasty, through the Yuan dynasty, to the death of Emperor T'ai-tsu, founder of the Ming dynasty. It is marked by a gracefulness in proportion and refinement in detail.

The Period of Rigidity lasted from the beginning of the fifteenth century to the end of the nineteenth century, or from the reign of Emperor Ch'eng-tsu (Yung-lo) of the Ming dynasty, who usurped his nephew's throne and moved the capital from Nanking to Peking, to the overthrow of the Ch'ing dynasty by the Republic. It is identified by a general rigidity, a clumsiness of proportion resulting from the excessive size of all horizontal members, and the decrease in size (in proportion to the building) of the *tou-kung*, hence the increase in number of the intermediate sets, which have by this time degenerated from their original structural function to mere ornaments.

This division of periods is, of course, fairly arbitrary. It is impossible to draw a line of demarcation to separate the imperceptible steps in the process of evolution. Thus a building of an early date may be found heralding a new style or feature, or in regions far from the cultural and political centers, a late structure may tenaciously cling to a bygone tradition. A generous margin must be allowed for the overlapping of the periods.

The Period of Vigor (ca. 850–1050)

Indirect Material Evidence

The Period of Vigor—only a few specimens of its final phase remain—must have had a brilliant past stretching considerably before the middle of the ninth century, perhaps as far back as the beginning of the T'ang dynasty, or early seventh century. As to the wooden buildings of this hypothetical first half of the period, we can only resort to contemporary graphic art for information. An incised relief on the tympanum over the west doorway of the Ta-yen T'a or Wild Goose Pagoda (701–704) in Sian, Shensi Province (fig. 22) , depicts in detail and with great accuracy the main hall of a Buddhist temple. It shows for the first time the brackets with outstretching arms, or *hua-kung*, used as cantilevers to support the great overhang of the eave. This does not imply that the *hua-kung* was not used previously; on the contrary, it must have been in use for quite a long time, perhaps even a few hundred years, before it developed into such an appropriate and mature feature and was depicted in a drawing representing what we may take to be a typical scene. The two tiers of jutting *hua-kung* are intersected by transverse *kung* and the inverted-V intermediate set is also used. The fin-shaped ornaments on the ridge ends and the bud-shaped ornaments on the hips are motifs that have since taken other forms. The only inaccuracy is the excessive slenderness of the columns, which are probably so presented in order not to obstruct the Buddhas and Bodhisattvas depicted within the building.

Other important sources of information are the paintings of the T'ang dynasty recovered by Sir Aurel Stein and Professor Paul Pelliot from the Caves of the Thousand Buddhas (fifth to tenth centuries) at Tun-huang, Kansu Province, now in the British Museum and the Louvre. In these paintings on silk and in the cave mural paintings are scenes representing the Paradise of the West, or the Happy Land of Amitabha. Many are set against highly architectural backgrounds of halls, towers, pavilions, pagodas, and the like. Here the *tou-kung* are seen not only with the outstretching *hua-kung*, but with the beveled ends of the slanting *ang*, a mode of construction employing the principle of the lever to make possible the far overhang of the eave, a device found in most of the monumental buildings of later periods. A great many other details can be gathered from these paintings (fig. 23) .

In some cave temples of the late T'ang dynasty in Szechuan Province, scenes of the same theme are shown in relief. The architecture depicted there is far simpler than that in the paintings, undoubtedly due to the limitations imposed by the medium. We can therefore infer, by comparing the sculptured and painted representations of this one theme, that the architecture depicted in the Han tombs and in the caves of the Wei, Ch'i, and Sui dynasties must have been a simplified version of a much more fully developed timber original.

Fo-kuang Ssu

The oldest wooden structure known today is the Main Hall of Fo-kuang Ssu, in the Wu-t'ai Mountains, Shansi Province (fig. 24) . [1] Built in 857, twelve years after the last nationwide persecution of Buddhism, it replaced a seven-bayed, three-storied, ninety-five-feet-high hall, which had housed a colossal statue of Mi-le (Maitreya) , also destroyed. The existing structure is a single-storied hall of seven bays. It is extremely impressive with its rigorous and robust proportions. The enormous *tou-kung* of four tiers of cantilevers—two tiers of *hua-kung* and two tiers of *ang*—measuring about half the height of the columns, with every piece of timber in the ensemble doing its share as a structural member, give the building an overwhelming dignity that is not found in later structures.

The interior manifests grace and elegance. Spanning the hypostyle columns are "crescent-moon beams", supported at either end by four tiers of *hua-kung* that transmit their load to the columns. In contrast to the severity of the exterior, every surface of the beams is curved. The sides are pulvinated, and the top and bottom are gently arched, giving an illusion of strength that would otherwise be lacking in a simple straight horizontal member.

The building's most significant feature, from the point of view of dating structural evolution, is the formation of the "truss" immediately under, and therefore supporting, the ridge. On the top tier of beams are placed a pair of *ch'a-shou* ("abutting arms") butting against each other to carry the ridge pole, while the "king post" is missing altogether. This is a rare survival of early construction practice. Similar construction is indicated in

---

[1] A smaller and simpler temple, Nan-chan Ssu, built in 782, was later discovered in the same region. See article in *Wen Wui* 1 (1954), 1:89. —Ed

the Stone shrine at the Tomb of Chu Wei (first century A.D.) , Chin-hsiang, Shantung Province (fig. 13) , and also in one of the Tun-huang paintings. In actual specimens, it is employed in the colonnade around the courtyard of Horyuji (seventh century) , Nara, Japan. The Fo-kuang Ssu example is the only one of its kind preserved in China and is never seen in later structures.

What makes the Main Hall even more of a treasure is the presence within of sculpture, painting, and calligraphy, all of the same date. On the large platform is a pantheon of nearly three dozen Buddhas and Bodhisattvas of colossal and heroic size. Of even more interest are the two humble, life-sized portrait statues, one of a woman, Ning Kung-yu, donor of the Hall, and the other of a monk, Yuan-ch'eng, the abbot who rebuilt the temple after the destruction of 845. On the undersides of the beams are inscriptions written with brush and ink, giving the names of the civil and military officials of the district at the time of the completion of the Hall, and also that of the donor. On one frieze a fresco of moderate dimensions, unmistakably T'ang in style, is preserved. In comparison the Sung painting in the next bay, dated 1122, itself a great treasure, is eclipsed. Thus in a single building are found examples of all four of the plastic arts of the T'ang dynasty. Any one of them would be proclaimed a national treasure; and the assemblage of all four is an unimaginable extravagance.

The next one hundred and twenty years is a blank period from which not a single existing wooden structure is known. Then there are two wooden structures in the Tun-huang caves, dated 976 and 980. Although they hardly deserve the name of real buildings, for they are merely porches screening the entrances of the caves, nevertheless they are rare examples of the architecture of the early Sung dynasty.

Two Buildings of Tu-le Ssu

Chronologically the next wooden structures are the magnificent Kuan-yin Ke, or Hall of Avalokitesvara, and the Main Gate, both of Tu-le Ssu, Chi Hsien, Hopei Province. They were built in 984, when that part of the country was under the occupation of the Liao Tartars. The Hall (fig. 25) is a two-storied structure with a mezzanine story. It houses a colossal clay statue of an eleven-headed Kuan-yin, about fifty-two feet in height, the largest of its kind in China. The two upper floors are built around the statue with a "well" in the middle, forming galleries around the figure's hips and chest. In terms of construction, then, the Hall is built of three tiers of "superposed orders", each complete with column and *tou-kung*. The proportions and details of the tou-kung show very little departure from that of the T'ang Main Hall in Fo-kuang Ssu. But here, in addition to the double-*kung*-double *ang* combination that is used on the upper story, there are, supporting the balcony and the lower eave, *tou-kung* without the *ang*, that is, with horizontal cantilevers only, similar to those depicted in the Ta-yen T'a engraving (fig. 22) . The interior is a grand exhibition of *tou-kung* of various combinations and in various positions, each arranged so as to do its share in supporting the entire structure. A straight girder, which has taken the place of the "crescent-moon beam" , supports the coffered ceiling of very small checkered squares. For carrying the load of the ridge pole, a "king post" —here a *chu-ju-chu* or "dwarf post" —is introduced in addition to the pair of *ch'a-shou* forming a simple truss. In due course the king post became the sole transmitter of the weight of the ridge pole to the beam, with the total elimination of the *ch'a-shou* truss arms. Thus the presence or absence of the *ch'a-shou* and their size, in proportion to that of the *chu-ju-chu* king post, often serve as convenient criteria for dating a building.

Another characteristic of this period, with rare exceptions, is that the interior or hypostyle columns are usually of the same height as the outer peristyle. The higher parts of the roof frame are supported by the stacking up of *tou-kung*. The columns are seldom lengthened, as is the practice of later ages, to reach for a closer contact with the upper members.

The Shan-men or Main Gate (fig. 26) of Tu-le Ssu is a small structure with very simple *tou-kung* under the eaves. The plan is typical of the Chinese gate building. Along the exterior longitudinal axis is a row of columns on which the doors are fixed. The interior treatment is the so-called open-flame construction, that is, there is no ceiling and all the structural members that support the roof are exposed. Here is exhibited an ingenious example of the carpenter's art, entirely structural in function but extremely decorative in appearance. This dual quality is the greatest virtue of the Chinese structural system.

Starting with these two buildings, we know of more than three dozen timber structures built during the next three hundred years, which include the Liao, Sung, and Chin dynasties. Though their number is small and their chronological distribution uneven, we can still arrange them into a continuity with no unbridgable gaps. About a dozen examples belong to what we call the Period of Vigor, and they are all found in the Liao domain in North China.

The Liao timber structure that follows chronologically the Kuan-yin Ke and Main Gate of Tu-le Ssu is the Main Hall of Feng-kuo Ssu (fig. 27) , I Hsien, Liaoning Province, built in 1020. In this large building outside the Great Wall, the intermediate bracket set is identical with the column set, which is a double-*kung*-double-*ang* combination, like that of the column sets of the Kuan-yin Ke. The corner set is further elaborated by placing an auxiliary *lu-tou* on the corner column at each side of the angle, i. e., combining two intermediate set with the corner set. This treatment, known as the *fu-chiao lu-tou*, or corner set with adjoining *lu-tou*, is quite common later but extremely rare in this early period (fig. 30i) .

The San-ta-shih Tien, or Hall of the Three Bodhisattvas (1025) , Kuang-chi Ssu, Pao-ti, Hopei Province, is very severe in external appearance but extremely elegant in interior treatment (fig. 28) . The bracketing is quite simple: only horizontal *hua-kung* are used. The interior is of the open-frame type, with no ceiling to conceal the structural features. This treatment gives the architect a grand opportunity to display his ingenuity in handling major carpentry as an artistic creation.

Two Groups at Ta-t'ung

In the city of Ta-t'ung, Shansi Province, there are two groups of buildings of great importance—the Hua-yen Ssu, inside the West Gate, and the Shan-hua Ssu, inside the South Gate. Both temples are recorded as having been founded in the T'ang dynasty, but

　　　　　　　　　　　　　图像中国建筑史

the existing structures date only from the middle of the Liao dynasty.

The Hua-yen Ssu was originally a large temple spreading over a broad acreage. It suffered much damage during the successive frontier wars, and now only three buildings of the Liao and Chin dynasties remain. Among these, the Library (figs. 29a-d) and its Side Hall are two Liao structures, the former dated 1038 and the latter probably constructed at the same time. (The Main Hall of Hua-yen Ssu, which is now the principal building of the so-called Upper Temple, belongs to the next period.)

The bracketing in the Library is similar to that in the Pao-ti structure, but the interior framing is all hidden by the ceiling. Lining the side and back walls are sutra shelves, highly architectural in treatment and exquisitely executed. They are invaluable not only as specimens of cabinetwork of that period, but also as examples of what the *Ying-tsao fa-shih* calls the *Pi-tsang*, or wall sutra cabinet. They are excellent models for the study of Liao construction. The Library also houses a group of superb Buddhas and Bodhisattvas.

The Side Hall of the Library is a modest structure with an overhanging gable roof. The bracketing is very simple. Especially noticeable is the use of a *t'i-mu* in the *lu-tou* as an auxiliary half-*kung* under the *hua-kung*, a peculiarity found only in a few buildings of the Liao period and never seen again (fig. 29e) .

One other unusual feature of the Hua-yen Ssu group is its orientation. In contrast to the orthodox practice of orienting principal buildings to the south, here the principal buildings face the east. This was an early custom of the Khitans (Liao Tartars) , who were originally sun worshippers and considered East the most noble of the four cardinal points.

The Shan-hua Ssu is a group that has retained much of its original arrangement (figs. 30a, b) . As the existing buildings show, the original party consisted of seven principal buildings on one main axis and two transverse axes. The whole group was originally surrounded by a continuous veranda, which is now discernible only by its foundations. Of the seven buildings only one, a flanking tower in the inner courtyard, was destroyed, and the existing ones are all early structures of either the Liao or the Chin dynasties. The connecting galleries and the monks' living quarters have vanished.

The Main Hall and the P'u-hsien Ke belong to the Period of Vigor. The Main Hall (figs. 30c-e) is a seven-bayed structure elevated on a high platform. It is flanked on its left and right by two aligned *to-tien*, or "ear halls" ; all three buildings face south. This aligned arrangement of ear-halls is in accordance with ancient tradition and is seldom seen in later buildings. The bracketing of this large structure is quite simple. One significant characteristic is the use of diagonal *kung* in the intermediate sets of the three central bays, which was first seen in the sutra cabinet in the Library of Hua-yen Ssu (fig. 29b) and was to become quite a fad in the Chin dynasty. The adobe wall is reinforced with horizontal courses of wooden "bones" , which are effective in preventing vertical cracks. This method was employed in some buildings of the thirteenth and fourteenth centuries, but it was not generally adopted.

The P'u-hsien Ke, or Hall of Samantabhadra (figs. 30f-h)is a two-storied building, very small, but constructed essentially in the same manner as the Kuan-yin Ke of Tu-le Ssu, The diagonal *kung* is again seen here.

These two buildings are of the Liao dynasty. Their exact date is unknown, but stylistically they may be placed in the middle of the eleventh century. The Front Hall and the Main Gate of the Shan-hua Ssu belong to the next period and will be discussed below.

The Fo-kung Ssu Wooden Pagoda

The Wooden Pagoda of Fo-kung Ssu, Ying Hsien, Shansi Province (fig. 31) , may be considered the grand finale of the Period of Vigor. It was built in 1056 and may have been a fairly common type at that time, for there are quite a few brick versions of it in the provinces of Hopei, Jehol, and Liaoning, then a part of the Liao domain.

The structure is octagonal in plan, with two rings of columns. Its five stories are built entirely of wood. It is similar to the Kuan-yin Ke of Tu-le Ssu in principle of construction: since each upper story is underpinned by a mezzanine story, it actually consists of nine tiers of superposed orders. The ground story is surrounded by a peristyle with a lean-to roof all the way around to give the effect of a double eave. The octagonal pyramidal roof of the top story is surmounted by a wrought-iron *sha*, or finial, the tip of which stands 183 feet [1] from the ground. There are altogether fifty-six different combinations of *tou-kung* in this structure, including all the kinds previously mentioned. The student of Chinese architecture can find no better collection for his studies.

The Period of Elegance (ca. 1000–1400)

Early Sung Characteristics

While the Liao Tartars were carrying on the rigorous traditions of the T'ang dynasty, the Sung architects were already introducing a style characterized by elegance and refinement. This period lasted for approximately four hundred years.

The most noticeable stylistic evolution is the gradual diminution of the *tou-kung*, which were reduced from generally about one third the height of the column to about one fourth by 1400 (fig. 32) . The intermediate set becomes relatively larger in size and more complicated in combination and, finally, not only assumes a shape exactly like that of the column set but even further complicates itself by the introduction of the diagonal *kung*. This intermediate set, because it supports no beams and does not rest on a column, becomes a burden to the architrave. In some cases, and also according to the *Ying-tsao fa-shih*, two intermediate sets are used in the central bay. When an *ang* is used in an intermediate set, it further complicates the structural problem by the presence of its tail extending upward at an angle. When used in a column set, the tail is usually held in position by the weight of the beam, but here the tails are cleverly used as intermediate supports for the purlin above. The *ang*'s disposition gives the architect a wonderful opportunity to display his ingenuity, resulting in various combinations of great interest. However, the structural function is never forgotten; the sets are always designed to do their share in upholding the whole structure and are seldom idle or merely decorative.

In the interior the columniation is often adjusted to suit a utilitarian purpose, such

[1] 疑为 220 feet。——译校注

as providing space to accommodate Buddhist images or worshippers. When columns are omitted in the hypostyle, not only is the plan affected but even the roof framing. (Specific examples will be discussed below.) In addition to this irregularity in columniation, the columns of the hypostyle are increased in height to form a more direct support to the beams above. Yet wherever a horizontal is joined to a vertical member, a simple set of *tou-kung* is always employed as a transition.

In buildings with ceilings, the framework hidden above is generally unfinished or rough. But a monumental building with open framework usually displays a maze of beams, lintels, ties, *tou* and *kung*, and tails of *ang*, interlacing, intersupporting and interdepending. Their disposition is an art in which the architects of this period excelled.

A Vanguard: Yu-hua Kung

Yu-hua Kung, the small hall of Yung-shou Ssu, near Yu-tz'u, Shansi Province, is a structure in which elegance is most skillfully blended with vigor (fig. 33) . Because of its early date (1008) it could be listed among the structures of the preceding period. But here are seen the first signs of the mellowness that will soon characterize the architecture of the Sung and Yuan dynasties. It is a specimen of the transition, embodying the virtues of both periods.

The hall's moderate size and dilapidated condition do not impress the casual onlooker. But its delightful beauty can never escape the eyes of the connoisseur. The *tou-kung* are extremely simple—single *hua-kung* and single *ang*—with the *shua-t'ou*, or head of the beam, shaped like an *ang* and installed in a slanting position to give the effect of two beaks. The set is proportionately large, less than one-third the column height, and the intermediate set is virtually eliminated. In the open- framework interior the exposed members, including the tail of the *ang*, are so neatly joined and fitted that they seem the outcome of an imperative logical sequence.

A Group in Cheng-ting

The Lung-hsing Ssu, Cheng-ting, Hopei Province, is a temple where a number of early Sung buildings is preserved. The Shan-men, or Main Gate, though in a fair state of preservation, was badly repaired in the eighteenth century, when small sets of *tou-kung* in the Ch'ing style were "smuggled" into the spaces between the large original Sung sets. The general aspect is ludicrous.

The Mo-ni Tien is the Main Hall of this temple (figs. 34a, b) . It is a double-eaved structure, nearly square in plan. On each of its four fronts is a projecting portico, with its gable-hip roof facing out. This treatment is often depicted in ancient paintings but rarely seen in existing structures. The *tou-kung* are very large and robust. Although there is only one intermediate set in each bay, the diagonal *kung*, which was a favorite of the Liao architects, is used. The very marked "crescendo" of the height of the column toward the corners has a soothing effect.

The Chuan-lun-tsang Tien (Hall of the Revolving Bookcase) is a library built around a revolving sutra cabinet (figs. 34c-g) . Here the spacing of the columns in the interior is modified to accommodate the sutra cabinet. This in turn affects the open-framework roof, in which the numerous structural members are ingeniously fitted together, not unlike a well-performed symphony, in which every part comes in at the precise time and place to achieve a perfect harmony.

The revolving sutra cabinet, an octagonal structure that turns on a pivot, is a unique specimen of its kind. In treatment it is a grand elaboration of the architectural elements, taking the form of a double-eaved pavilion. The lower eave follows the shape of the octagon; the upper eave is circular. Both eaves are treated with elaborate *tou-kung*. Because this piece of cabinetwork is executed in close accordance with the specifications in the *Ying-tsao fa-shih*, it is an invaluable example of Sung construction. Unfortunately it was brutally abused by soldiers when the temple was used as a barracks and had been left in a pitiable state when last seen by the author in 1933.

The Chin-tz'u Group

Another important group of the early Sung dynasty is the Sheng Mu Miao,Temple of the Saintly Mother, at Chin-tz'u, near T'aiyuan, Shansi Province (fig. 35) . The group consists of a double-eaved Main Hall, reached by a bridge over a rectangular pool, in front of which is a Front Hall and a *p'ai-lou*, and still farther to the front a terrace with four cast-iron guardians. The two halls and the bridge were built in the T'ien-sheng period (1023– 1031) . Except for the decorative paintings, these structures are well preserved and have been little spoiled by later restorations.

A prominent characteristic of the *tou-kung* of the Lung-hsing Ssu and Chin-tz'u groups is the shape of the beak of the *ang* (fig. 36) . In earlier buildings, the beak is a simple bevel, rectangular in cross section, which makes an angle of approximately 25 degrees with the underside of the *ang*. This is known in the *Ying-tsao fa-shih* as the *p'i-chu ang*, or "split bamboo *ang*." But in these two later groups, the beveled portion is scooped and pulvinated, resulting in a cross section with a rounded top, known as the *ch'ing-mien ang*, or "lute-face *ang*." [1] This is the orthodox shape of the beak of an *ang* from the time of the *Ying-tsao fa-shih* till the present day, though later the pulvination is reduced to a mere beveling of the edges. In these two groups the second treatment—the pulvinated straight bevel—characterizes all the *ang*. It is found only in Sung structures of the early eleventh, and possibly the end of the tenth centuries, for by the time of the publication of the *Ying-tsao fa-shih* (1103) , the third treatment was already standardized. The Mo-ni Tien and the Chuan-lun-tsang Tien of Lung-hsing Ssu, and the Wen-shu Tien of Fo-kuang Ssu (see below) , are the other surviving buildings that show this second characteristic and are, therefore, of approximately the same date.

Another peculiarity of this short period is the shaping of the *shua-t'ou* in exactly the same manner as the beak of the *ang* (fig. 37) . In earlier examples it is usually shaped with either a straight vertical, cut or a simple *p'ei-chu* bevel, while later it assumes the form of a "dragon head" or "grasshopper head" (*ma-cha t'ou*) . Shaping it exactly like the beak of an *ang* and installing it at a similar angle to create the impression of an extra *ang* are,

---

[1] 疑为or "lute-face ang"。 The third treatment is to bend the bevel into a concavity, still pulvinated, like a segment of the inner surface of a tubular ring. This is the orthodox shape...——译校注

however, treatments found only in structures built at the turn of the eleventh century.

At this point we may also bring up the *p'u-p'ai fang*, the plate immediately above the lintel (fig. 38). The plate and the lintel are combined to form a T-shaped cross section. Its earliest appearance is in the Yu-hua Kung of Yu-tz'u, Shansi Province (1008), and it is also found in the Cheng-ting and Chin-tz'u groups, but it is comparatively rare in such early structures. Even in the *Ying-tsao fa-shih*, its use is limited to the *p'ing-tso*, or mezzanine story. But by about 1150 the *p'u-p'ai fang* had become an integral part of every building; an architrave without a plate was now an exception.

Another feature of the *tou-kung* that made its debut in the Chin-tz'u structures is the false *ang*. Here the *hua-kung*, a horizontal member, is shaped on the outer end into a beak like that of an *ang*. Thus the decorative effect of the beak is obtained without using the slanting member itself. The *hua-kung* becomes merely an "applied ornament," a sign of degeneration. This treatment eventually became standard in the Ming and Ch'ing dynasties, and its appearance at this very early date is a disheartening portent of the departure from structural integrity that was to come.

A Unique Construction: Wen-shu Tien

The Wen-shu Tien, or Hall of Manjusri, side hall to the T'ang structure in Fo-kuang Ssu, is a seven-bayed structure, with overhanging gables, of unimposing facade (fig. 39). Its *tou-kung* are similar in treatment to those of Lung-hsing Ssu and Chin-tz'u. It is, however, interesting as a unique example of interior framing. Because of its peculiar columniation, only two hypostyle columns are used in the central bay at the rear, so that the span between them is lengthened to three bays, about forty-six feet from center to center of the columns. No timber of average size could span this length, so a compound framework resembling a modern "queen-post" truss was introduced. However, the construction, which is really no truss in the structural sense, did not live up to the task expected of it by its designer, and an auxiliary support had to be installed in later years.

Examples Contemporary with the *Ying-tsao fa-shih*

Of the existing Sung buildings, the one closest to the *Ying-tsao fa-shih* in date is a small hall, the Ch'u-tsu An of Shao-lin Ssu, in the Sung Mountains, Honan Province (fig. 40). It is a square structure of three bays. The stone columns are octagonal; one bears the date 1125, only twenty-two years after the publication of the book. The general structural features abide quite closely to the rules, and the *tou-kung* are executed in almost complete obedience to them. A minor feature revealing a Sung characteristic is the treatment of the sides of the ramps: the triangular walls are treated with a series of sunken panels, one inside another, exactly as specified in the *Ying-tsao fa-shih*.

Farther to the north, in the domain then under the Chin Tartars, are also several buildings of approximately the same date. The Main Hall of Ching-t'u Ssu (fig. 41), Ying Hsien, Shansi Province, built in 1124, is a year closer in date to the *Ying-tsao fa-shih* than is the Ch'u-tsu An. In spite of the political and geographical distance, the general proportion of this building adheres quite closely to the Sung rules. Its ceiling, treated with the *t'ien-kung lou-ke*, or "heavenly palaces" motif, as given in the *Ying-tsao fa-shih*, is a magnificent piece of cabinetworker's art.

The Front Hall and the Main Gate of Shan-hua Ssu, Ta-t'ung, Shansi Province (fig. 42), in the domain of the Chin Tartars, were built between 1128 and 1143. In the Front Hall the intermediate sets of *tou-kung* with diagonal *kung*, which had previously appeared in the Mo-ni Tien of ung-hsing Ssu, the Wooden Pagoda of Ying Hsien, and elsewhere, have developed into formidable jumbles of intersecting *kung* and *tou*, imposing themselves as a terrific load on the architrave. The *p'u-p'ai fang*, the plate resting on the lintel, has become thicker than the few examples of earlier dates. In the open-framework interior the *tou-kung* are very effectively employed.

The Shan-men, or Main Gate, of Shan-hua Ssu is perhaps the most pretentious of its kind (fig. 43). The five-bay structure is certainly larger than many a main hall of smaller temples. Its *tou-kung* are simple, and the diagonal *kung* does not appear. The "crescent-moon beam" is used, which is very rare for northern China.

Contemporary with the Chin Tartars was the Southern Sung dynasty. So far only one wooden structure of this period is known in the south: the Main Hall, San-ch'ing Tien, of the Taoist temple Hsuan-miao Kuan, Soochow, Kiangsu Province. Although it was built in 1170, not even three-quarters of a century after the *Ying-tsao fashih*, it has lost much of the vigor of that time and appears much more ornate than contemporary and even later structures in the north. The smallness of the *tou-kung* in proportion to the hall as a whole is especially noticeable.

The Last Phase of the Period of Elegance

Quite a number of specimens in both the north and the south survive from the last century and a half of the Period of Elegance. During this time the *tou-kung* underwent many important changes. One was the widespread use of the false *ang*, which first appeared in the Chin-tz'u group. Another was the increasing size of the *shua-t'ou*, the "grasshopper-head" outer beam-end, on the column set (see fig. 37) . As the *tou-kung* shrank through the ages, the *shua-t'ou*, when cut to the ancient proportion of one *ts'ai*, proved too flimsy for its structural function; so, proportionally, it had to become larger than one *ts'ai*. In order to receive it, the *hua-kung* underneath also had to become wider. (By the time of the *Kung-ch'eng tso-fa tse-li* of 1734, the *shua-t'ou* on the column set had swollen in width to 40 *fen*, or four *tou-k'ou*, four times the Sung proportion. The width of the lowest tier of *hua-kung* had become twice its ancient width, from ten *fen* to 20 *fen*, or two *tou-k'ou*.)

The Yang-ho Lou of Cheng-ting, Hopei Province (ca. 1250) is an excellent example of the last phase of the Period of Elegance. It is a seven-bayed structure, a sort of belvedere on a high masonry terrace, which is penetrated by two vaulted tunnels like a city gate, standing as a monumental feature over the main thoroughfare of the city (fig. 44) . The *tou-kung* appear to be double *ang*, but in reality both tiers of *ang* on the column set are false, and on the intermediate set one is false and one real. The architrave is pulvinated and shaped at both ends like the crescent-moon beam, but of course it is not arched. This treatment is found in a few other buildings of the Yuan dynasty.

Among other examples are the Te-ning Tien, Main Hall of the Pei-Yueh Miao (1270), Ch'u-yang, Hopei Province (fig. 45) , and the Ming-ying-wang Tien of the Temple of the Dragon King at Kuang-sheng Ssu (ca. 1320) , Chao-ch'eng, Shansi Province (fig. 46) . The latter is decorated with a mural painting representing a theatrical performance of the Yuan dynasty, dated 1324, a rare example of a secular subject used in the decoration of a religious building.

The Upper Temple and the Lower Temple of Kuang-sheng Ssu, Chao-ch'eng, Shansi Province, are two interesting groups of which the buildings are constructed in a most unorthodox manner (fig. 47) . In these late Yuan of early Ming structures are found enormous *ang*, so employed that, in some cases, even beams are dispensed with. Such construction is also, found in a few other buildings in southern Shansi, for example, the Temple of Confucius of Lin-fen. But it is not seen in other localities or other periods, therefore it may be a purely local characteristic.

The Main Hall of Yen-fu Ssu (1324–1327), Hsuan-p'ing, Chekiang Province (fig. 48) , is a rare specimen of a Yuan building in the Lower Yangtze region and south of the river. The open-framework structure is a masterpiece of intricate carpentry. Though characteristically Yuan, it has a suavity and lightness in prominent contrast to the heavier construction of the north.

There are also a few buildings of the Yuan dynasty in the southwestern province of Yunnan. It is interesting to note that the remote regions, though slow to catch up with the general proportions of structures in the eastern cultural centers, were quite apt in copying details of the contemporary period. The Yunnan buildings are twelfth or thirteenth century in general proportion but fourteenth century in treatment of details.

The Period of Rigidity (ca. 1400–1912)

With the founding of the capital at Peking at the beginning of the fifteenth century, there appeared, principally in the official architecture of the court, a style of marked departure from the tradition of the Sung and Yuan dynasties. The change is very abrupt, as if some overwhelming force had turned the minds of the builders toward an entirely new sense of proportion. Even during the reign of Emperor Hung-wu (1368–1398), founder of the Ming dynasty, buildings were still constructed in the Yuan style. A few good examples of this last flicker of the Period of Elegance are the City Gates (1372) and the Drum Tower (probably also 1372) of Ta-t'ung, Shansi Province, and the Fei-lai Tien of Fei-lai Ssu (1391), O-mei, Szechuan Province.

The sudden change in the proportion of the *tou-kung* in the structures of the new capital is at once noticeable (see fig. 32). Measuring about one-half or one-third the height of the column in the Sung dynasty, they are in the Ming suddenly shrunken to one-fifth. The number of intermediate set, never exceeding two in the twelfth century, has now increased to four or six, and later to even seven or eight. Instead of sharing the duty of lifting the overhanging eave, which they can no longer do by means of nimble "jumps," they impose themselves as a load on the architrave. The architrave, which used to be more of a tie than a girder in the previous periods, must now be enlarged to carry this additional load. The *p'u-p'ai fang* no longer forms a "T" with the architrave, but is flush with it, or even narrower in some cases, for the small *lu-tou* of a shrunken set does not require a plate of excessive width.

In the *tou-kung* itself other great changes have taken place. The *ang*, with its long tail, has no place in a column set with a disproportionately large beam resting directly on it. The false *ang* is therefore always employed where the effect of *ang* is desired. But in the intermediate sets in the interior, the tail of the *ang* is profusely employed, not as a structural member but as a decorative motif. The tails are embellished with an exuberance of carved and applied ornaments, especially the *san-fu yun*, which was originally a simple wing-shaped transverse member occasionally applied on a *hua-kung* with a "stolen heart" in the Sung dynasty, but is now elaborated into "a cloud cluster" and attached to the tail. These tails, however, are no longer the upper parts of the slanting *ang* whose lower (outer) ends are shaped into beaks. The beak is now the extension of the *hua-kung*, resulting in the false *ang*. The tail has become the extension of a horizontal member, such as the *shua-t'ou* or the *ch'en-fang-tou* (the small tie beam above the *hua-kung*) . When these horizontal members are extended by means of a long slanting tail, they become something like a hockey stick in appearance. They are no longer the lever arms that hold up the purlin, but a burden to be supported by an auxiliary lintel. The *tou-kung*, except in the column sets, is therefore by now pure ornament.

In the disposition of beams forming the roof support, the *tou-kung* is dispensed with altogether and the beams, now larger, rest directly on columns or struts. The purlins are supported directly on the ends of the beams, without the assistance of *kung* or the bracing of *ch'a-shou* or *t'o-chiao*. The now sturdy king post supports the ridge purlin alone.

Columniation is so regular that the plan becomes a checkerboard. After about 1400 a column is seldom omitted to make room for a utilitarian requirement.

Emperor Yung-lo's Tomb

The earliest surviving example of this style of architecture is the Sacrificial Hall of the Tomb of Emperor Ch'eng-tsu (Yung-lo) , built between 1403 and 1424, at the Thirteen Ming Tombs, Ch'ang-p'ing, Hopei Province. All the subsequent emperors and empresses of the Ming dynasty were buried in this neighborhood, but Yung-lo's tomb was the grandest and dominated at the center.

The Sacrificial Hall (fig. 49) , which is the principal building in the tomb compound, is a nine-bayed, double-eaved hall, elevated on three tiers of marble terraces. It is almost an exact replica of the Feng-t'ien Tien, Emperor Yung-lo's audience hall in the Imperial Palaces in Peking (see below) . The *tou-kung* are extremely small in proportion, with tails of unprecedented length. As many as eight intermediate sets are used, all of them purely ornamental. Both the small size of the *tou-kung* and the large number of intermediate sets are uncommon in structures of this early phase of the period. The general effect of the building, however, is most impressive.

Ming Buildings among the Peking Palaces

The Imperial Palaces of the Ming dynasty at Peking were rebuilt on the ruins of the sacked Yuan capital, the Ta-tu that Marco Polo visited. Since its rebuilding five and a half centuries ago, it has remained little changed in general parti till the present day. Although the Huang-chi Tien (called Feng-t'ien Tien at the beginning of the Ming dynasty and replaced by the T'ai-ho Tien in the Ch'ing dynasty) , the principal audience hall, was destroyed at the fall of the Ming dynasty, there are still a considerable number of Ming buildings in the Forbidden City. Of these the Hsiang Tien, or Sacrificial Hall, of She-chi T'an—in the Central Park—is the oldest, built in 1421 (fig. 50) . Here the *tou-kung* still measures about two-sevenths the column height, and only six intermediate sets are installed in the central bay, four in the side bays. An interesting group in the Palaces is the Imperial Ancestral Temple, or T'ai Miao, rebuilt principally in 1545 (fig. 51) .

The Chien-chi Tien (renamed Pao-ho Tien) , the last of the "Three Great Halls," is a structure dating from 1615, rebuilt after a fire (fig. 52) . [1] It fortunately escaped destruction when the two halls in front of it were burned down at the fall of the dynasty in 1644 and again in a conflagration in 1679. There is little difference between it and buildings of the period of the *Kung-ch'eng tso-fa tse-li* (1734) , either in general proportions or in details. It would have been impossible to ascertain its earlier date had it not been for the inscriptions on the structural members above the ceiling; on each the Ming designation "Chien-chi Tien" and the position of the member is written with brush and ink.

The Library in the Temple of Confucius (1504) , Ch'ü-fu, Shantung Province (fig. 53) , is an interesting example of a two-storied building in the official style of the Ming dynasty. The influence of this rigid style, however, does not seem to have spread far beyond Peking, or rather, beyond the buildings erected by imperial order and according to official regulations. Elsewhere in the empire architects were given more freedom than in the service of the court. Buildings clinging more or less to older traditions are found all over the country. The T'ien-tsun Tien of Wen-ch'ang Kung, Chih-t'ung (middle

---

〔1〕疑为 "is a structure dating from 1615 (fig. 52)"。——译校注

Ming), and the group at Chiu-feng Ssu (1443 and later) , Peng-hsi, Szechuan Province, are excellent examples.

Ch'ing Buildings in Peking

The architecture of the Ch'ing dynasty (1644–1912) is a mere continuation of the Ming tradition. With the publication of the *Kung-ch'eng tso-fa tse-li* in 1734, innovation was stifled. In the structures built for the emperors throughout the 268 years of the regime there is a uniformity that no modern totalitarian state could achieve. Most of the buildings in the Imperial Palaces in the Forbidden City and the Imperial Tombs and numerous temples in and around Peking are products of this school (figs. 54–56) .

The most outstanding examples are, of course, the Imperial Palaces (fig. 57) . As individual buildings, especially from a structural point of view, they are unremarkable. But as a composition of the grand plan, there is nothing comparable to it in the world: it is a grand plan on the grandest scale. The mere idea of laying out an axis nearly two miles in length from south to north, with an endless series of avenues, courts, bridges, gates, colonnades, terraces, pavilions, halls, palaces, balanced with perfect symmetry on both sides, and all built in exactly the same fashion, in strict accordance with the *Kung-ch'eng tso-fa tse-li*, is a most appropriate expression of the Son of Heaven and of a powerful empire. Here, the uniformity induced by the strict rules turns out to be more of a merit than a defect. Without such rigid restrictions, dignity and grandeur of such magnitude could never have been achieved.

There is, however, a grave drawback in this composition. It seems that the planner totally overlooked, or we may venture to say that he was incapable of coping with, the secondary or transverse axes. Even in the grouping of buildings immediately flanking those on the main axis, the relation between the axes is not always comfortable. Thus we find that almost all the groups in the Forbidden City, especially the "apartments" on both sides of the series on the principal axis, though well balanced within the four walls of each group—each with its north and south axis parallel to the main axis of the

Forbidden City—bear no definite relation to it transversally. Although this emphasis on one axis is in fact an outstanding characteristic of all Chinese planning, as is shown in all the temples and residences throughout China. It is almost unbelievable that a planner who paid so much attention to symmetry on one axis should have ignored so completely the other.

The T'ai-ho Tien (Hall of Supreme Harmony) , the principal audience hall of the palaces and the focal point of the entire Forbidden City, is the grandest individual building (fig. 58) . With its six rows of twelve columns each, forming a hypostyle hall eleven bays in length and five in depth, double-eaved and hip-roofed, it is the largest single premodern building in China. The seventy-two columns, arranged in monotonous regularity that shows any quality except ingenuity, are still a most impressive sight. The hall stands on a low marble platform, which in turn is elevated by three tiers of balustraded terraces, profusely decorated with exquisite carvings. The structure dates back no farther than 1697, built to replace the one destroyed by a fire in 1679.

The *tou-kung* of this gigantic hall are extremely small in proportion—less than one-sixth of the column height. As many as eight intermediate sets are used in one bay. From a distance the presence of *tou-kung* is hardly noticeable. The walls, columns, doors, and windows are painted vermilion, the *tou-kung* and architraves are blue and green accented with gold. The entire structure is crowned by a roof of glazed yellow tiles, glistening like gold in the bright sunshine against the blue northern sky. Surmounting the white marble terraces that seem to vibrate with their exuberant carvings, the great hall is a rare feast for the eye, an unforgettable apparition of grandeur, dignity, and beauty.

As an aside, it may be mentioned that the T'ai-ho Tien is equipped with a most unusual "fire proofing system." In the dark attic under the roof, above the ceiling of the central bay, is a tablet bearing the names and the magic words of both Buddhist and Taoist gods who have control over fire, wind, and the thunderbolt. In front of this tablet stand an incense burner, a pair of candles, and a pair of *ling-chih*, a fungus symbolizing Taoist immortality. So far, the precaution seems to have worked.

Another famous building of the Ch'ing dynasty is the Ch'i-nien Tien of the Temple of Heaven (fig. 59). The hall is circular in plan, with three decks of eaves, the uppermost of which forms a conical roof. The glazing of the tiles is blue, the color symbolic of heaven. This beautiful structure is also elevated on three tiers of balustraded marble terraces. The present structure was built in 1890 to replace the one burned down the year before.

A typical Buddhist temple of the Ming and Ch'ing dynasties is the Hu-kuo Ssu in Peking. Founded in the Yuan dynasty, the temple was almost completely rebuilt in the Ch'ing period. The buildings are grouped to allow circulation from one courtyard to another around the sides of the buildings. The drum tower and bell tower standing on the left and right of the forecourt are erected only in temples. One hall in the rear of the compound, a Yuan structure that is now a complete ruin, is an interesting example of the use of timber reinforcements in adobe walls.

A number of mosques were also built in the Ming and Ch'ing dynasties. But except for details in the interior decoration, they show no essential difference from other buildings.

The Temple of Confucius, Ch'ü-fu, Shantung

A building almost contemporary with the *Kung-ch'eng tso-fa tse-li* but departing considerably from its rules is the Ta-ch'eng Tien, Main Hall of the Temple of Confucius (1730), Ch'ü-fu, Shantung Province (fig. 60) . Although it is impressive for its carved marble columns, in general proportion it has a certain awkwardness not found in other Ch'ing buildings, whether erected by imperial mandate or otherwise.

The Temple of Confucius, which has been in state custody ever since the Han dynasty, is the only group of buildings in China that has an unbroken history for over two thousand years. The present compound is a large group that occupies the entire center of the town of Ch'ü-fu; its present plan was laid out in the Sung dynasty. The numerous buildings within its walls are an interesting assortment of structures of various periods. The oldest is a stela pavilion of the Chin dynasty, 1196[1], the latest is a building erected

---

[1] 疑为"1195"。——译校注

in 1933, and examples from the three intervening ynasties—Yuan, Ming, and Ch'ing—are all present.

Construction Methods in the South

Away from the influence of government regulations, sometimes not even far from Peking, a certain lightness in treatment gives buildings an appearance of liveliness. This manifestation is most pronounced in the provinces south of the Yangtze River. The difference is explained not only by the warmer climate but also by the greater technical agility developed by the southerners. In the milder south there is no necessity for the heavy masonry walls and thick roof construction so imperative for shutting out the severe cold of the north. Walls of plastered laths and roofs with tiles laid directly on rafters without even sheathing are common. Timbers are generally smaller, and roofs often have high up-turned corners that are very playful in spirit. When this tendency is carried to excess, it often results in faulty construction and frivolous ornamentation, at the expense of restraint and purity, two virtues that a good architect can hardly afford to be without.

Residential Architecture

The domestic architecture of the province of Yunnan seems to have struck a happy medium in combining the agility of the south with the restraint of the north. It shows a certain freedom in planning not found in other parts of China. Units of various sizes and functions are joined together, seemingly very casually, with interesting intersections of roofs and romantic fenestrations with "eyebrow eaves" and balconies, often attaining a high degree of picturesqueness in mass composition.

Rural architecture built of local materials, as in the mountain village near Wu-yi, Chekiang Province, is typical of southern China (fig. 61) . But in the northern regions of loess plateaus, cave dwellings dug into the loess cliffs are still quite common. In the mountains along the Burma Road in western Yunnan are unusual log cabins, not unlike those of Scandinavia and America in construction. But there is something so essentially

Chinese about them, especially in the treatment of the roofing and the porch, that it is hard to deny that an architecture is always permeated with the spirit of its people, as is expressed here even in these out-of-the-way, casual, and modest little cabins (fig. 62) .

Buddhist Pagodas

As an architectural monument, giving expression and accent to the landscape of China, nothing figures more prominently than the *t'a*, or pagoda. From its first appearance till the present day, the Chinese pagoda has remained essentially "a multi-storied tower surmounted by a pile of metal discs, " as we have quoted. It is the happy combination of two principal components: the indigenous "multi-storied tower" and the Indian stupa, the "pile of metal discs." By combinations of the two components, the Chinese pagoda may be classified into four principal types: One-storied, multi-storied, multi-eaved, and stupa. Whatever its size or type, a *t'a* marks the burial site of a Buddhist relic or the tomb of a Buddhist monk.

We know from literary sources such as the citation above, from the replicas in the Caves of Yun-kang (fig. 17) , Hsiang-t'ang Shan, and Lung-men, and from existing examples in Japan, that the early pagoda was an indigenous multi-storied tower, square in plan, constructed of timber, and surmounted by a stupa, called the *sha*, or spire. It did not take long for the builders to recognize the advantages of brick and stone for a monument of this kind, and the masonry pagoda soon made its appearance and eventually outlasted its wooden prototypes. With a single exception, the Wooden Pagoda of Ying Hsien (fig. 31) , all the pagodas in China today are masonry structures.

The masonry pagoda's evolution (fig. 63) can be divided broadly into three periods: the Period of Simplicity, or the Period of the Square Plan (ca. 500–900) ; the Period of Elaboration, or the Period of the Octagonal Plan (ca.1000–1300) ; and the Period of Variety (ca. 1280–1912) . As with our classification of wooden buildings, generous margins must be allowed for the overlapping of, and departure from, stylistic and temporal characteristics.

The Period of Simplicity (ca. 500–900)

The first period may be marked from the early sixth century to the end of the ninth century, which includes the Wei, Ch'i, Sui, and T'ang dynasties. The most prominent characteristics, with a few rare exceptions, are the square plan and the one-cell hollow construction; that is, the pagoda is built like a shell, with no internal horizontal or vertical masonry divisions (although there may be wooden floors and stairs) . The structure resembles a modern factory smokestack, with the top closed and covered. Three of the four types of pagoda made their appearance in these centuries. The stupa, although an Indian prototype that might be expected in China in this early period, is curiously absent, though there is every reason to believe that it was not unknown. There are many examples of the other three types.

One-storied Pagodas

Pagodas of the one-storied type, with a single exception, are all tomb monuments to deceased monks. They are usually small and are more like shrines than pagodas as the latter term is generally understood. This type is depicted in great numbers in the reliefs of the Yun-kang Caves. It is characterized by a square cella with an arched door-way on one wall, and it is crowned by one or two strings of eaves, or cornices, and is surmounted by a *sha.*

The oldest stone pagoda in China, and the oldest and most important surviving one-storied type, is the Ssu-men T'a (544 A. D.)[1] of Shen-t'ung Ssu, near Tsinan, Shantung Province (fig. 64a) . But it is by no means typical because it is neither a tomb monument

[1] 疑为 "611"。——译校注

nor a hollow shell. It is a single-storied, pavilionlike, ashlar structure, square in plan, with a central core surrounded by a gallery, and an arched doorway in each of the four sides. It is the only pagoda of this period with such an interior arrangement, which became the common characteristic of pagodas in the tenth century and thereafter.

Typical examples of the many one-storied tomb pagodas are the Tomb Pagoda of Hui-ch'ung (ca. 627–649) in Ling-yen Ssu, Ch'ang-ch'ing, Shantung Province (fig. 64b) , and the Tomb Pagoda of T'ung-kuang (771) in Shao-lin Ssu, in the Sung Mountains, Honan Province (fig. 64c) .

A unique specimen of great architectural importance is the Tomb Pagoda of Ching-tsang at Hui-shan Ssu, Teng-feng, also in the Sung Mountains of Honan (figs. 64d, e) . Built soon after the death of the monk in 746, it is a small, octagonal, single-storied, pavilionlike brick structure on a high moulding-decorated base, or *shu-mi-tso*. The exterior is treated with engaged columns at the corners, *tou-kung*, false windows, and other elements. The *tou-kung* are similar to those of the Yun-kang and T'ien-lung Shan Caves except for the projection of a *shua-tou*, intersecting the *kung* in the *lu-tou*. An inverted-V intermediate set is placed on the architrave on each face of the octagon. The cella is characteristic of the period, but the octagonal plan and the *shu-mi-tso* appear here for the first time. They are to become the two salient characteristics of pagodas after the latter half of the tenth century. Their introduction in this mid-eighth-century structure presages an important change in the evolution of the pagoda.

Multi-storied Pagodas

The multi-storied pagoda, the type most commonly represented in the Yun-kang Caves both in reliefs and in the round, is the superposition of a number of one-storied pagodas. As the structure rises, each story diminishes slightly in both height and breadth. The exterior of the extant brick specimens is often treated with pilasters with a very slight reveal and with simple sets of *tou-kung*, in imitation of the timber construction of the time. The most famous example of this type is the Wild Goose Pagoda, the Ta-yen T'a (figs. 65a, b) of Tzu-en Ssu, Sian, Shensi Province. The original structure, consecrated by the great master Hsuan-tsang in the mid-seventh century, was soon destroyed, and the present structure was built between 701 and 704. Its construction is the typical hollow shell with timber floors and stairways. The exterior is treated with extremely slender pilasters in very delicate relief, in great contrast to the vigorous massive bulk of the structure. A single *tou* is used above each pilaster, and the intermediate set is eliminated. Each side of each story is perforated by an arched doorway. Above the west doorway of the first story is found the valuable engraving depicting a T'ang hall of timber construction (fig. 22) .

Two other notable examples of pagodas of this type are the Hsiang-chi Ssu T'a and the Hsuan-tsang T'a, both in the neighborhood of Sian. The Hsiang-chi Ssu T'a (669) [1] is quite similar to the Ta-yen T'a in general parti and in the treatment of exterior walls. But here a single *tou* is used as an intermediate set, and blank windows are represented on the walls (fig. 65c) . The Hsuan-tsang T'a in Hsingchiao Ssu (681)[2] is a small structure of five stories. It may or may not be the tomb monument to the great master. The ground-story treatment is astylar, that is, without columns, while the four upper stories are treated with pilasters. It is similar to the Tomb Pagoda of Ching-tsang in the treatment of *tou-kung* in which a *shua-t'ou* projects from the *lu-tou*, but here the *shua-t'ou* is vertically cut. This structure, which antedates the Ching-tsang T'a by nearly seventy years, is the earliest specimen showing such a treatment (fig. 65d) .

An extremely unusual multi-storied pagoda is the so-called Tsu-shih T'a, or Founder's Tomb Pagoda, of Fo-kuang Ssu, in the Wu-t'ai Mountains, Shansi Province (fig. 65e) . Standing a mere few paces to the south of the T'ang Main Hall of 857, it is a two-storied hexagonal structure. The lower story has a cornice of lotus petals and corbelled brick courses, resting on a string of *tou*. (The lotus-petal motif is employed in the treatment of all three tiers of cornices.) The upper story is raised on a *p'ing-tso*, the "mezzanine" or balcony support. This *p'ing-tso* is a *shu-mi-tso*, with "legs" that divide the *shu-yao*, or dado, into small compartments. This treatment of the *shu-mi-so*, though a minor feature, is one of the most characteristic of the middle and later sixth century, and can be fairly taken as a token

〔1〕疑为 "681"。——译校注
〔2〕疑为 "669"。——译校注

of its date. The *p'ing-tso* is an important feature in the structure of a multi-storied building in timber, as we have seen, and its appearance here is significant (fig. 65f) .

The upper story is more architecturally treated. The corners are decorated with engaged columns with lotus blossoms at both ends and at the middle, showing an unmistakable Indian influence. The front wall has an arched doorway with a flame-shaped extrados or archivolt, and on each of the two adjacent walls is a window with vertical bars. Above one window an inverted-V intermediate set is still preserved, painted on the plastered wall in dark red (fig. 65g) . The drawing has an energetic tension that also characterizes the draperies of the Buddhist sculpture and the calligraphy of the Northern Wei and Ch'i dynasties.

The exact date of this pagoda is not known. But judging from the *shu-mi-tso*, the flame-shaped archivolt, the lotus-blossom column, the inverted-V intermediate set, and other stylistic evidence, it must have been built in the latter part of the sixth century.

Multi-eaved Pagodas

The multi-eaved type is characterized by a high principal story, usually without a base, and with many courses of eaves, or cornices, above. The eaves are usually odd in number, seldom fewer than five and rarely exceeding thirteen. The combined height of the eaves is usually about twice that of the principal story. Because it is customary in China to designate the number of stories in a pagoda by the number of eaves, this type of pagoda is called "so-many-storied," which is misleading. Structurally, or architecturally, the eaves of this type are superposed directly one above another, with hardly any intermediate space. Therefore we shall use the designation "multi-eaved pagoda."

A masterpiece, although anything but typical, is the Pagoda of Sung-yueh Ssu, Teng-feng, in the Sung Mountains, Honan Province, built in 520. Indeed, it is a unique structure with its fifteen courses of eaves and twelve-sided plan. The general parti consists of an unusually high base, on which is elevated the principal story. This is decorated at each corner with an engaged column topped by a lotus-bud capital, a motif of Indian origin. Every side except the four that open with an arched doorway is treated with a niche containing a relief of a one-storied pagoda on a "lion pedestal." The courses of eaves diminish in a gentle parabolic curve, presenting an unusually graceful silhouette. The entire structure is a hollow shell of brick masonry, octagonal on the interior (figs. 66a, b) . Because the original wooden floors and stairs have vanished the interior appears not unlike that of an elevator shaft.

The typical multi-eaved pagodas of this period are all square in plan and stand directly on the ground; the principal story is generally plain, without any decoration. The Pagodas of Yung-t'ai Ssu and of Fa-wang Ssu, both early eighth century and both in the Sung Mountains, Teng-feng, Honan Province, are the best examples (figs. 66c, d) .

The Hsiao-yen T'a (684) [1] of Chien-fu Ssu, Sian, Shensi Province, is a pagoda of this type but with windows on the narrow bands of wall separating the eaves (figs. 66e, f) . The eaves, however, especially the higher ones, do not correspond with the floors and therefore do not express the interior divisions. The difference in height between the principal story and those of the upper "stories" is so marked, in contrast to the more regular progression of stories of the multi-storied type, that the pagoda must properly be classified as multi-eaved. The Pagoda of Fo-t'u Ssu (820?) and the Ch'ien-hsun T'a (ca.850) , both in Tali, Yunnan Province, are also multi-eaved (figs. 66g-j) .

A number of stone pagodas of the T'ang dynasty belong to the multi-eaved type. They are usually small structures, rarely more than twenty-five or thirty feet high. The eaves are rather thin slabs carved into steps in imitation of the corbelled courses of the brick pagoda. The doorway on the principal story is usually carved as an arch with a flame-shaped archivolt and is flanked by two *dvrapalas* or guardian kings. Typical examples are the four stone pagodas of Yun-chu Ssu (711–722)[2], Fang-shan, Hopei Province, which stand on the four corners of a large terrace as satellites to a large, later pagoda in the center (figs. 66k, l) . This grouping of five pagodas on a common base will become quite usual in the Ming and Ch'ing dynasties.

[1] 疑为 "709"。——译校注
[2] 疑为 "711–727"。——译校注

The Stupa

At about the end of the tenth century there appeared the stupa, a pagoda type that is more Indian than any we have seen. Tombs, in the form of Indian stupas with nearly hemispherical bodies, are frequently depicted in the mural paintings in the Tun-huang caves. Actual specimens of this date, however, though by no means rare in Chinese Turkestan, are very uncommon in China proper. One example is the Stupa at Fo-kuang Ssu. The stupa type eventually established a strong foothold in China and is discussed below under the Period of Variety.

The Period of Elaboration (ca. 1000–1300)

The Period of Elaboration may be tentatively set from the end of the tenth to the end of the thirteenth centuries, that is, covering the Five Dynasties and the Northern and Southern Sung dynasties, as well as the Liao and Chin dynasties. The pagoda of this period is characterized by an octagonal plan and by the introduction of both horizontal and vertical masonry interior divisions in the form of galleries and builtin stairs. These divisions are a wide departure from the old hollow shell and give an entirely changed aspect to the interior. But the idea is not new, for it had already appeared once in the middle of the sixth century, in the Ssu-men T'a of Shen-t'ung Ssu (fig. 64a) .

The octagonal pagoda which first appeared in the Tomb Pagoda of Ching-tsang in 746 (figs. 64d, e) was the first "pa-go-da" in the real sense of the term. The origin of this peculiar word has always been a mystery. Perhaps the most plausible explanation is that it is simply the southern pronunciation of the three characters *pa-chiao-t'a* "pa-ko-t'a, " meaning "eight-cornered pagoda." The word "pagoda" instead of "t'a" is deliberately used in this book because it is accepted in all the European languages as the name for such a monument. The very fact that the word finds its way into almost every European dictionary as the name for the Chinese *t'a* may refleet the popularity of the octagonal plan at the opening of Western contact.

The external aspect of the pagoda of this period is characterized by the ever-more faithful imitation of timber construction. Columns, architraves, elaborate and complicated sets of *tou-kung*, eaves with rafters, doors, windows, balconies with railings, and the like all make their appearance on the brick pagoda. Thus the multi-storied and multi-eaved pagodas of this period look quite different from their earlier prototypes.

One-storied Pagodas

The one-storied pagoda, which was very popular in the Period of Simplicity, became a rarity after the end of the T'ang dynasty. The few that remain from the twelfth century are still square in plan. The shrinelike pagoda is raised on a base of *shu-mi-tso*, a treatment not found in the T'ang dynasty examples. The entrance, which opened into the cella in the previous period, is now closed with doors, on which appear rows of "nail-heads" and knockers. Typical examples are the tombs of P'u-t'ung (1121), of Hsing-chun (926) (fig. 64f) , and of Hsi-t'ang (1167) , all in the Shao-lin Ssu in the Sung Mountains, Honan Province.

Multi-storied Pagodas

As the one-storied pagoda gradually disappeared, changes also took place in the taller structures. The octagonal plan became the norm, the square plan the exception. The wooden floors and stairs that used to divide and connect the stories gave way to masonry floors and stairs. The architects were timid at first: the pagodas were built almost as solid masonry, bored through with narrow "tunnels" for galleries and stairways. As the builders gained in experience and audacity, the masonry was increasingly lightened by enlarging the galleries on each floor, until the structure became virtually a brick-masonry core with a semidetached shell. The two were bonded at intervals by arched or corbelled floors.

The multi-storied pagodas of this period may be divided into two subtypes: the "timber-frame subtype" and the "astylar subtype." The former may further be subdivided into the Liao or northern style, and the Sung or southern style.

The Liao timber-frame subtype may best be described as the rendition in brick of such structures as the Wooden Pagoda (1056) of Fo-kung Ssu, Ying Hsien, Shansi Province (fig. 31) , which is the only surviving wooden pagoda in China. There are typical examples in Hopei Province—the Twin Pagodas of Yun-chu Ssu, Cho Hsien [1] , built in 1090 (figs. 67b, c) , and the Ch'ien-fo T'a of I Hsien (fig. 67a)— and a few similar structures in the provinces of Jehol and Liaoning. Their faithful adherence to the wooden form is obvious. The only difference in proportion is the narrow overhang of the eaves, due to the limitations of the material.

A pagoda belonging to the timber-frame subtype, but unique among all the pagodas in China, is the Hua T'a, or Flowery Pagoda of Kuang-hui Ssu, Cheng-ting, Hopei Province, so called because of its flowery appearance (figs. 67d, e) . An octagonal brick structure of very fanciful plan and startling mass composition, it is treated with timber-frame forms on the exterior and surmounted by a conical "spire" profusely decorated with the most fantastic motifs. Four hexagonal one-storied pagodas are attached to four sides of the ground story. This peculiar plan is probably a modified version of the arrangement of the Pagoda Group of Yun-chu Ssu, Fang-shan. The exact date of the Flowery Pagoda is unknown, but, judging by the stylistic characteristics of the timber construction imitated, it is probably of the late twelfth or early thirteenth century.

An unusual structure in the northern timber-frame subtype is the Pei T'a, or North Pagoda (fig. 67f) , the central feature of the Yun-chu Ssu group, which stands amid the four earlier stone pagodas of T'ang date. This pagoda is only two stories high, apparently an unfinished structure of the Ying Hsien Wooden Pagoda type. It is surmounted by a large stupa with a hemispherical "belly" and a large conical "neck" . The lower two stories are undoubtedly of the Liao dynasty; the upper part may be later.

The southern Sung timber-frame subtype is quite common in the lower Yangtze region. The earliest examples are three small stone pagodas located in Hangchow, Chekiang Province. These miniature pagodas (about thirty feet tall) , which are really

[1] 疑为 the Twin pagodas of Cho Hsien。——译校注

Dhanari columns in pagoda form, are carved with great dexterity and are no doubt very faithful copies of wooden structures of the time. Two examples are the Twin Pagodas (960)in front of the Main Hall of Ling-yin Ssu(figs. 68a, b) , and the third is the Pai T'a in the railroad yard at the Cha-k'ou Station.

However, true pagodas typical of this southern style are exemplified in the Twin Pagodas of Lo-han Yuan(982) , Soochow, Kiangsu Province, and the Tiger Hill Pagoda in the same city(figs. 68c-g) . Compared to their northern counterparts they are characterized by a marked slenderness in general proportion, which is further emphasized by the tall, slim metal *sha*, or spire, and the absence of the high *shu-mi-tso* base. In detail the columns are shorter but of more delicate entasis. The *tou-kung* are simpler, due to the absence of the diagonal *kung* and to the "stolen heart" on the fifirst "jump" . The rafters are merely suggested by the chevron courses, called *ling-chiao ya-tzu*, in the narrow corbelled cornice. Thus the eave overhang is very slight, giving the pagoda a silhouette entirely different from that of the northern structures.

The two ashlar pagodas at Chin-chiang (Ch'uan-chou) , Fukien Province, the Chen-kuo T'a and the Jen-shou T'a (1228–1247) , are stone versions of the southern timber-frame subtype. Their affinity to the wooden form is obvious. Because almost all masonry pagodas are brick structures, these two examples in stone are valuable rarities.

The astylar subtype, that is, the structures that are not treated with columns on the facade, is primarily a Northern Sung product. Existing specimens are mostly found in the provinces of Honan, Hopei, and Shantung, and less commonly in other parts of China. These pagodas are characterized by the complete absence of columns or pilasters on the walls. The *tou-kung*, however, is often used; there are a number of examples with cornices built up by corbelled courses of brick. The *tou-kung* may not necessarily appear in distinct separate sets, but they are often arranged in a row, producing a band of vibrating light and shade under the eave. In most cases the architrave is shown above the wall to receive the *tou-kung*.

Typical of this subtype with *tou-kung* treatment are the P'i-chih T'a of Ling-yen Ssu

(fig. 69a) , Ch'ang-ch'ing, Shantung Province, and the Pagoda of Shen-kuo Ssu, Hsiu-wu, Honan Province, both of the late eleventh century. An excellent example of the corbelled eave kind is the Liao-ti T'a of K'ai-yuan Ssu (1001) , Ting Hsien, Hopei Province (fig. 69b) . The northwest section [1] of this structure has completely fallen off, exposing the internal construction like a model of a cross section made especially for the convenience of the student of Chinese architecture (fig. 69c) . Pagodas of this kind are also found in the southwest, such as the Pai T'a of Pao-en Ssu (ca. 1155) , Ta-chu, and another at Lu Hsien, both in the province of Szechuan.

The Sung dynasty introduced the use of terra-cotta as facing for pagodas. The so called "Iron Pagoda" of Yu-kuo Ssu (1041) [2], K'aifeng, Honan Province, is a notable northern example (fifigs. 69d, e) . The structure is a multi-storied pagoda with a very slight indication of columns at the corners, and is faced with terra-cotta tiles. Glazed in brown, they produce an effect like that of oxidized iron, from which the structure got its popular name.

Pagodas actually made of iron are, however, not uncommon in the Sung dynasty. They are usually small and very slender because of the nature of the material—cast iron. Like the stone pagodas of Hangchow, these miniature pagodas are really Dhanari columns in the form of pagodas. Examples are found in the Yu-ch'uan Ssu, Tang-yang, Hupei Province, and in Tsi-ning, Shantung Province.

Multi-eaved Pagodas

After the fall of the T'ang dynasty the multi-eaved pagoda became almost exclusively an expression of the Liao and Chin Tartars. It is the type commonly seen in North China today. Treated with timber forms, it now presents an aspect quite different from its predecessors in the Period of Simplicity. The octagonal plan, with a few rare exceptions, is universally adopted. But the structure has become a solid pile of masonry, no longer to be entered and ascended. The whole structure is invariably raised on a high *shu-mi-tso* base, which in turn is sometimes elevated on a broad, low platform. The

---

[1] 疑为 The northeast section。——译校注
[2] 疑为 "1049"。——译校注

principal story is treated with engaged columns at the corners, architraves, and blind doors and windows on the walls. The successive tiers of eaves are usually supported by *tou-kung*, accurately rendered in brick. But the corbelled eave is also common, and sometimes both treatments are combined in one pagoda by limiting the *tou-kung* to the lowest eave only.

A typical example is the well-known pagoda of T'ien-ning Ssu, Peking (fig. 70a) . Above the *shu-mi-tso* a *p'ing-tso* of lotus petals further elevates the structure. The doors are flanked by guardian kings, and the windows by standing Bodhisattvas. The building is of the eleventh century but somewhat freely repaired in later periods.

A good example with corbelled upper eaves is the Pagoda of T'ai-ning Ssu, I Hsien, Hopei Province. The Ch'ing T'a of Lin-chi Ssu (1185) , Cheng-ting, Hopei Province, is a smaller structure of the same type. The Tomb Pagoda of Chen-chi, Po-lin Ssu (1228) , Chao Hsien, Hopei Province, is a similar structure with a slight modification, the indication of a very low story under each of the upper eaves (fig. 70b) . This departure from the norm may be considered a compromise between the multi-eaved and multi-storied types.

Occasionally multi-eaved pagodas of this period still cling to the traditions of the Period of Simplicity, such as the square plan. A good example is the Ta T'a, or Big Pagoda, of Feng-huang Shan, Ch'ao-yang, Liaoning Province, probably of the thirteenth century. The pagoda of Pai-ma Ssu, near Loyang, Honan Province, built in 1175 when that region was under the Chin Tartars, is a square pagoda with thirteen tiers of cornices (fig. 70c) . The plan and the general proportions are obviously in the T'ang manner. But the structure is a solid and is raised on a *shu-mi-tso*, both characteristics not found in the previous period.

Another example, the Pai T'a of I-pin, Szechuan Province, ca. 1102–1109, is absolutely T'ang in character if judged only by its appearance. But the interior arrangement, with a celia in the core, surrounded by a continuous gallery and stairs, is characteristic of its own time (fig. 70d) .

Square pagodas of the T'ang style are quite numerous in the southwestern provinces, especially in Yunnan where they were built in the old manner as late as the Ch'ing dynasty.

Dhanari Columns

One peculiar Buddhist monument that originated some time in the T'ang dynasty and became very popular in the Period of Elaboration is the *ching-chuang*, or Dhanari column, which is also called *ching-t'a*, or Dhanari pagoda. The distinction depends upon the similarity of the edifice to a real pagoda. These monuments vary widely in "architecturesqueness," from a simple octagonal column set on a *shu-mi-tso* base and crowned by a sort of cornice or umbrella, to the shape of a real pagoda, though always much smaller.

The Dhanari column in front of the Main Hall of Fo-kuang Ssu, dated 857 and bearing the name of Ning Kung-yu, donor of the Hall, is a typical simple example (fig. 71a). The column in Feng-ch'ung Ssu, Hsing-t'ang, Hopei Province, probably of the twelfth century, is characteristic of numerous *ching-chuang* of the Sung and Chin dynasties. The *ching-chuang* of Chao Hsien, Hopei Province, highly elaborate in treatment and very graceful in proportions, is the largest of its kind known (fig. 71b). It is also attributed to the early Sung dynasty, probably the eleventh century. Another interesting specimen is the thirteenth-century Dhanari column of Ti-tsang An, Kunming, Yunnan Province. The whole is conceived more as a piece of sculpture than architecture. Monuments of this kind seem to have lost their popularity with the fall of the Sung dynasty.

The Period of Variety (ca.1280–1912)

We shall call the period from the founding of the Yuan dynasty in 1279 to the fall of the Ch'ing dynasty in 1912 the Period of Variety. Its first innovation was the sudden popularity of the "bottle-shaped" pagoda (the Tibetan version of the Indian stupa), which followed the introduction of the Lamaist sect of Buddhism by the conquering Mongols. This type had been faintly heralded about three centuries earlier in the Stupa of Fo-kuang Ssu. In the Chin dynasty it had been adopted in various modified forms as a tomb monument for monks. Finally it became firmly established in the Yuan dynasty.

The second innovation was the formulation in the Ming dynasty of the *chin-kang pao-tso t'a*, or the clustering of five pagodas on a common high terrace. Again, this arrangement had already been suggested in the five pagodas of Yun-chu Ssu of Fang Shan, Hopei (fig. 66k) , in the early eighth century. But it lay dormant for more than seven hundred years, until the latter part of the fifteenth century. Though it never became popular enough to be found throughout China, the existing examples constitute a definite type.

Large numbers of the more conventional types of pagoda were built in the Ming and Ch'ing dynasties, not always at the service of Buddhism. They were often erected as devices for the control or rectification of certain defects in *feng-shui* (literally, "wind and water") , a geomancy based on the influence over human destinies of the elements of nature, especially topographical peculiarities and orientation. A common expression of this belief is the *wen-fang* t'a, the tower that brings good luck to those taking literary examinations. They are seen throughout southern China, where they are usually erected on an elevated terrain to the south or southeast of many a walled town.

Multi-storied Pagodas

With the end of the Chin dynasty in 1234, the multi-eaved pagoda suddenly fell out of popularity, and the multi-storied pagoda became the dominant type. Its outstanding characteristics during the Ming dynasty were its increased slenderness and its squatter division of stories. The general contour shows less entasis and, in some cases, more tapering in a harsh straight line. The eaves became mere rings, with a narrow overhang and very flimsy *tou-kung* (if used at all) , reflecting their diminished proportion in the wooden originals of the day. Examples of this type are numerous. The Pagoda of Ching-yang, Shensi Province, an early-sixteenth-century structure, is quite reminiscent of the Period of Elaboration. The pagoda of Ling-yen Ssu, Fen-yang, Shansi Province, built in 1549, is a most typical Ming pagoda. The Twin Pagodas of Yung-chao Ssu (fig. 72a) , Taiyuan, Shansi Province, built in 1595, have a comparatively greater overhang of the eaves, which cast deeper shadows for a more accentuated appearance than is normally found in pagodas of the Ming dynasty.

The Fei-hung T'a of Kuang-sheng Ssu, Chao-ch'eng, Shansi Province, built in 1515, is a peculiar example (fig. 72b) . The thirteen stories diminish in a strong taper, without the slightest entasis, presenting the appearance of an awkwardly proportioned octagonal pyramid. The awkwardness is emphasized by the addition of an excessively wide timber gallery around the ground story. The exterior is burdened by a profusion of ornaments in yellow and green glazed terra-cotta, while *tou-kung* and lotus petals are employed as supports for the eaves of alternate stories. In construction the building is virtually a solid pile of masonry, bored through by a zigzag stairway tunnel that is most ingeniously constructed for the total elimination of landings (fig. 72c) .

A monument of more or less the same category, and equally distasteful in appearance, is the square pagoda of Ta-yun Ssu, Lin-fen, Shansi Province, built in 1651 (fig. 72d) . The five-story square structure is surmounted by an octagonal "lantern" , which is a most novel way of interpreting the *sha*. In the ground-story cella is a colossal head of Buddha, about twenty feet high, resting directly on the ground. This iconographic treatment is itself as unorthodox as the design of the pagoda that houses it.

A few other multi-storied pagodas of the Ch'ing dynasty may be mentioned. The Pagoda of Hsin-chiang, Shansi Province, is of the astylar subtype. Though adhering much to the tradition of the last period, it is characterized by entasis which is carried too high, resulting in a spikelike contour. The Pagoda of Feng-sheng Ssu, Chin-tz'u, near T'ai-yuan, Shansi Province, is of the northern timber-frame subtype, with a pleasing appearance not unlike that of the Liao structures (fig. 72e) . The Pei T'a, or North Pagoda, of Chin-hua, Chekiang Province, is a representative southern example (fig. 72f) .

Multi-eaved Pagodas

With the fall of the Chin dynasty, construction of multi-eaved pagodas virtually ceased. Besides the two pairs of Twin Pagodas in Peking, which are rather small and were originally Yuan structures but almost completely rebuilt during the Ch'ing dynasty, the only "full-size" multi-eaved pagoda we know from the Yuan dynasty is the one at T'ien-ning Ssu, Anyang, Honan Province (fig. 73a) [1]. This structure as it now stands, with its five decks of eaves or stories, is evidently unfinished, like so many of the towers of Gothic cathedrals in France. Like the tomb Pagoda of Chen-chi, Po-lin Ssu, Chao Hsien, Hopei Province (fig. 70b) , there is a very low story under each tier of eaves. Therefore, it is not even multi-eaved in the strict sense, but appears so only in general effect. It is terminated by a *sha* in the form of a stupa typical of the Ch'ing style. The timber-form details represented on the facades of the principal story reflect very faithfully the wooden construction of the day. The pagoda is unusual in plan as well as in exterior appearance, for it is not solid like similar structures of the Liao and Chin dynasties. Except for the ground story, which has a stairway around the celia, the whole is a hollow shell like the T'ang pagodas. This plan is rarely seen in later ages.

Though no longer favored as a full-size edifice, the multi-eaved pagoda became quite popular as a tomb monument in the Yuan dynasty. Two excellent small examples are the Tomb Pagoda of Hung-tz'u Po-hua Ta-shih and the Tomb Pagoda of Hsu-chao (ca. 1290) , both in Hsing-t'ai, Hopei Province. The latter is hexagonal in plan and is surmounted by a hemispherical stupa (fig. 73b) . In the southwestern province of Yunnan, where the influence of the cultural centers was slow to permeate, the multi-eaved pagoda was still built under the Mongols, but usually in the T'ang style with a square plan and corbelled cornices.

The only multi-eaved pagoda of the Ming dynasty is the Tz'u-shou Ssu Pagoda at Pa-li-chuang, Peking (fig. 73c) [2]. Built in 1578, it was unquestionably inspired by the nearby T'ien-ning Ssu Pagoda (fig. 70a) , whose general proportion it follows closely. But in detail it shows all the traits of the late Ming dynasty, such as the slight projection of the

[1] 疑为 "fig. 73c"。——译校注
[2] 疑为 "fig. 73a"。——译校注

moldings of the *shu-mi-tso*, the comparatively low main story, the arched windows, the double architrave, and the small *tou-kung*.

A small pagoda, erected as a garden feature in the Yu-ch'uan Shan (Jade Fountain Park) near Peking in the eighteenth century, shows an interesting innovation of the Ch'ing dynasty. It may be considered a combination of the multi-storied and multi-eaved types. Of its three stories, the lower two are double-eaved, while the top story is triple-eaved, presenting a not unpleasant effect. The plan, though octagonal, is not equilateral, and it may be better described as a square with splayed corners. The whole structure is faced with glazed terra-cotta, and timber construction is very faithfully imitated on the facades.

Lamaist Stupas

As we have seen, the Lamaist stupa was heralded in the hemispherical tomb in Fo-kuang Ssu in the latter part of the tenth century and was adopted in certain tomb features in the Chin dynasty, but it did not formally appear as a monumental edifice until the Yuan dynasty. Its appearance assumed a new form: a bottle-shaped super-structure on a high base. The base is usually a *shu-mi-tso* of one or two tiers, the plan of which we shall call, for want of a better name, a square with "folded in corners," or with projections on the four sides. The "bottle" consists of a "belly" and a "neck," which is often crowned by an "umbrella."

The prototype is the Pai T'a of Miao-ying Ssu, Peking (fig. 74a) . It was built in 1271 by order of Kublai Khan, who deliberately had the existing Liao pagoda destroyed in order to replace it with the present structure. It is distinguished from later pagodas of its type by its sturdy proportions, the almost vertical sides of its "belly, " and the truncated-cone "bottle-neck."

The outstanding bottle-shaped pagoda of the Ming dynasty is the Pagoda of T'a-yuan Ssu (1577) , in the Wu-t'ai Mountains, Shansi Province (fig. 74b) . Here the bottom of the "bottle" is slightly diminished, while the top of the "neck" is enlarged. The general aspect is more slender than that of the Pai T'a of Peking.

A stupa of the Ming dynasty, whose exact date is uncertain, is the Pagoda of Shan-kuo Ssu, Tai Hsien, Shansi Province. The *shu-mi-tso* base, with its simple, bold moldings, is circular in plan and much larger than usual in proportion to the superstructure. The recession of the upper tier of *shu-mi-tso* is greater than usual. The "belly" has a soft, gentle contour, while the "neck" is pinched around its bottom as it rises from its own *shu-mi-tso* base. The general effect is one of steadiness and elegance, and the stupa may be considered the best proportioned bottle-shaped pagoda in China.

In time the Lamaist stupa became more slender, especially in the "neck." Two typical examples are the Pei T'a of Yung-an Ssu (1651) in the Pei Hai (North Lake) Park, Peking (fig. 74c) , and the stupa surmounting the archway at the entrance to the ruined site of Fa-hai Ssu (1660) , near the Hunting Park, Ching-yi Yuan, in the Western Hills, Peking. The two structures, though differing greatly in size, are almost identical in general proportion. In both, the high *shu-mi-tso* base is simplified to only one tier. The moldings above the *shu-mi-tso* and around the bottom of the "bottle," which formerly presented the cross section of a cyma, have now become taller, and at Fa-hai Ssu, they have become an additional stepped base, following the profile of the *shu-mi-tso* base below. The niche on the "belly," has now become a standardized feature. The "neck" has become almost cylindrical, with only a slight tapering, and is very slender in proportion to the "belly."

A few monuments of the same type are found in the neighborhood of Shenyang (Mukden) , Liaoning Province, relics of the early Ch'ing dynasty. They are distinguished by their very large "bellies" and broad bases. Various modifications of this type, usually small and executed in bronze, are used as ornamental features in temple grounds. Those in front of the Main Hall of Hsien-ch'ing Ssu in the Wu-t'ai Mountains, Shansi Province, are typical.

Five-pagoda Clusters

The one important contribution of the Ming dynasty to the art of pagoda building is the definite formulation of the *chin-kang pao-tso t'a*, or the "Vajra-based pagoda" the grouping of five pagodas on a common base. We have seen the first suggestion of such an arrangement the Yun-chu Ssu group of 711–722 (fig. 66k). In the middle of the T'ang dynasty (late eighth century), a more exuberant manifestation appears in the Pagoda of Ch'iu-t'a Ssu near Ch'ang-ch'ing, Shantung Province, where nine small square multi-eaved pagodas are clustered on top of an octagonal one-storied pagoda. In the Chin dynasty the idea takes a curious form in the Flowery Pagoda of Cheng-ting (fig. 67d). However, it was not until the T'ien-chun period (1457–1464) of the Ming dynasty, with the erection of the Chin-kang Pao-tso T'a at Miao-chan Ssu, near Kunming, Yunnan Province, that the type was definitely formed. This group is composed of five Lamaist stupas on a common terrace or base. The slender "neck", refined by a slight entasis, is unusual for the stupa type, while the niche in the "belly" is unusual for this early date. The terrace is penetrated by two intersecting barrel vaults, but no access is provided for ascending it.

The most important five-pagoda cluster is the Chin-kang Pao-tso T'a of Cheng-chueh Ssu, popularly known as the Wu-t'a Ssu, or Five-Pagoda Temple, outside the west wall of Peking (fig. 75a). It was built in 1473. The monumental is divided into five stories and treated with continuous rows of cornices, suggestive of a Tibetan temple building. The south facade is penetrated by an archway that opens onto the stairway leading to the top of the terrace. The pagodas themselves are of the square, multi-eaved type. A pavilion is placed in the center on the front of the terrace, forming a "penthouse" that covers the upper entrance of the stairway.

Another important example of this type is the Chin-kang Pao-tso T'a of Pi-yun Ssu, in the Western Hills, Peking (figs. 75b, c). In this group, erected in 1747, the arrangement of the Wu-t'a Ssu group has been further elaborated. Two additional bottle-shaped stupas are placed in front of the five square, multi-eaved pagodas. The pavilion in front, here standing between and behind the two stupas, is itself a smaller terrace, upon which the same theme of the five pagodas is repeated. The whole ensemble is further elevated by the two tiers of high terraces of rubble stone masonry on which it rests.

The group in the Yellow Temple, also of the eighteenth century, outside the north wall of Peking, is much smaller in scale. The Lamaist stupa in the center is fantastic in shape, while the four corner pagodas are octagonal and multi-storied. The group is placed on a low base on a low platform and preceded by a *p'ai-lou*.

Other Masonry Structures

The Chinese builder never cultivated a real knowledge of brick and stone as primary structural media for ordinary use. They were either reserved for structures that were not intimately involved with everyday life, such as city and garden walls, bridges, gateways, or tombs, or assigned to secondary functions, such as curtain walls of the walls coming up to the window sills of timber-framed buildings. Masonry structures, therefore, occupy a position in no sense comparable to that in European architecture.

Tombs

The oldest remains of vaulted construction are the brick tombs of the Han dynasty, which are found in great number. The subterranean structures are never architecturally treated and are thus of little architectural interest. Above ground the tomb or tumulus is usually preceded by an avenue, flanked at the entrance by a pair of *ch'üeh* (piers) , then by figures of attendants, guards, and animals, and terminated by a shrine in front of the tumulus. The *ch'üeh* and the shrine are the only items of architectural interest, for the figures are of more importance to the student of sculpture than to the student of architecture. The tombs of the Six Dynasties and the T'ang dynasty are of even less interest to us, as all the remains are sculptural.

The incidental discovery of a few twelfth-century tombs in the neighborhood of I-pin and Nan-hsi, Szechuan Province, has revealed the high degree of architectural treatment of tombs of the Southern Sung dynasty. These small ashlar funeral chambers are given architectural elements that closely imitate wooden construction of the time. The end facing the entrance is invariably in the form of a pair of half-open doors from which emerges a female figure(fig. 76a) . Similar tombs have not yet been found in other parts of China, so whether this is a purely local type or not is open to question.

There are many tombs from the Ming and Ch'ing dynasties, the most important of which are the imperial tombs. The aboveground structures are essentially temples, placed in front of the tumulus. It would be quite appropriate to call such groups "tomb temples." Their only important distinguishing feature is perhaps the *fang-ch'eng*, or "square bastion," surmounted by the *ming-lou*, or "radiant tower, " which marks the entrance to the *ti-kung*, or "underground palace." The Tomb of the Ming emperor Yung-lo(figs. 76b, c) , at Ch'ang-p'ing, is the most magnificent of the imperial tombs at Ch'ang-p'ing(Ming) , I Hsien(Ch'ing) , and Hsing-lung(Ch'ing) , all in Hopei.

The Tomb of the Ch'ing emperor Chia-ch'ing (1820) , at I Hsien, is a typical underground palace. From beneath the square bastion a passage leads to a series of gates, doors, passageways, antechambers, and finally to the "golden chamber," where the coffin of the emperor is laid. These "palaces" are vaulted, finished in marble, decorated with carvings, and generally roofed with yellow glazed tiles, exactly like a surface structure, though they were covered over with the ramped earth and lime-concrete that formed the tumulus. In some cases, however, they were not roofed, usually by will of the deceased emperor who wished to set an example of frugality. Drawings from the files, or rather from the wastepaper dumps, of the Lei family, hereditary architects to the imperial household of the Ch'ing dynasty for several hundred years, furnish interesting information about these "forbidden structures"(fig.76d) .

Vaulted Buildings

Except in the province of Shansi, the use of masonry for aboveground structures is extremely rare. A common type of house in Shansi is the vaulted house, which usually consists of three to seven barrel vaults placed side by side and parallel to each other. Penetrations cut as doorways allow circulation from vault to vault. The vaults are usually elliptical of parabolic in cross section, and the open ends are filled in with windows atop "window-sill walls." The spandrels are filled in to form a flat top, reached by an exterior stairway.

A vaulted building, called the *wu-liang tien* or "beamless hall, " is occasionally used as the main hall in a temple. The temple of Yung-chao Ssu (1597) , where the Twin Pagodas are also located, T'aiyuan, Shansi Province, is the best example (figs. 77a, b) . Here the barrel vaulting runs lengthwise, with penetrations for doors and windows like those of ordinary Shansi dwellings. But the exterior is treated with columns, architraves, *tou-kung*, and other elements. Similar structures are also found in the Hsien-ch'ing Ssu, in the Wu-t'ai Mountains, Shansi Province, and at Soochow, Kiangsu Province (figs. 77c, d).

The practice of architecturally treating the exterior of a brick vaulted structure to resemble a conventional wood-frame main hall did not exist before the latter half of the Ming dynasty, though it had been common on pagodas. The treatment is analogous to the application of the classical orders in Renaissance buildings in Europe. It is interesting to note that with the arrival of Matteo Ricci at Nanking in 1587 and the beginning of Jesuit influence on Chinese culture, the vaulted buildings already existing in Shansi prepared a fertile ground. The proximity in date of the arrival of the Jesuits in China and the appearance of the *wu-liang tien* may not be entirely coincidental.

In the neighborhood of Peking are a few *wu-liang tien* of the Ch'ing dynasty (fig. 77e) . Much larger than the Ming structures in Shansi, they are astylar in treatment. The "columns" are "hidden" in their massive walls: only the tops of the columns are indicated with architraves and *tou-kung*. These features are all in glazed terra-cotta. It is difficult to understand why these very impressive and monumental buildings were not more widely adopted.

Bridges

The earliest bridges in China were of wood, and pontoon bridges were used for crossing wide rivers. It was not until the fourth century A.D. that the first arched bridge was recorded in literature.

The oldest arched bridge in China today is the An-chi Ch'iao, popularly known as the Great Stone Bridge (figs. 78a, b) , near Chao Hsien, Hopei Province. It is an open spandrel bridge with two smaller arches at each end of the principal arch, which is segmental. The bridge is 115 feet long between the points where the two ends now emerge from the banks of the river. The clear span, if excavated from the banks, would be considerably longer. But our attempts to find its spring line by excavation at the abutments failed when we reached the water level seven feet below the dry riverbed.

The bridge is the work of the master builder Li Ch'un of the Sui dynasty (581– 618) . The principal arch is composed of twenty-eight separate strings of arches, placed side by side. The builder was fully aware of the danger of the strings falling apart, as is evidenced by his narrowing the top of the bridge toward the crown to give an inward incline to the exterior arches, thus counteracting the eccentric tendency of the separate rings. However, his foresight and ingenuity did not succeed against time and the elements. Five rings on the west side fell off in the sixteenth century (restored not long afterward) , and three on the east fell in the eighteenth century.

A smaller bridge of similar design stands outside the west gate of the same city, Chao Hsien, known as the Little Stone Bridge (fig. 78c) . It was built by Pao Ch'ien-erh at the end of the twelfth century, undoubtedly a copy of the Great Stone Bridge but only about half its length. The balustrades, dated 1507, show details closely akin to earlier wooden balustrades, and are interesting examples of the transition from the early more "wooden"

图像中国建筑史

forms to the type prevalent in the Ming and Ch'ing dynasties. The panels in the lower portion of the slabs show reliefs of interesting design.

As with buildings, bridge design also became standardized in the Ch'ing dynasty (fig. 78d) . There are numerous bridges of the official style in the neighborhood of Peking. The most famous is the Lu-kou Ch'iao, known as the Marco Polo Bridge, which was originally built at the end of the twelfth century during the Ming-ch'ang period of the Chin dynasty. But the present structure of eleven arches, about 1000 feet long overall, is an eighteenth-century replacement after the original bridge was swept away by a flood. It is the historical site where the Japanese army suddenly attacked the Chinese garrison in a "maneuver" in 1937, known as the Lu-kou Ch'iao Incident, which led to the Sino-Japanese war (fig. 78e) .

Arched bridges in southern China are generally lighter in construction than those of the official style. The thirteen-arched bridge of Chin-hua, Chekiang Province, is a superb example (fig. 78f) . Bridges of stone piers, spanned by timber girders and with covered roadways, are very common in the south. A typical example is the Bridge at Fu-min, Yunnan Province. Bridges with pillars built of stone drums, also spanned by timber, are found over the rivers Ch'an and Pa, near Sian, Shensi Province (figs. 78g, h) . Bridges of enormous stone slabs are not uncommon in the province of Fukien. The suspension bridge is widely used in Szechuan, Kweichow, Yunnan, and Sikang (figs. 78i, j) .

Terraces

The *t'ai*, or terrace, was commonly used for recreation as early as the Shang dynasty. Besides many records of such constructions in ancient chronicles, a great number of ruins of early terraces still exist in northern China. Notable among them are the more than a dozen *t'ai*, at the site of the capital of the Kingdom of Yen (second half of the third century B.C.) , near I Hsien, Hopei Province. But they are now nothing but large loess platforms, about twenty or thirty feet high, whose original appearance can in no way be reconstructed.

A terrace used for religious purposes is called a *t'an*, or altar. The most illustrious is the Yuan-ch'iu of the Temple of Heaven (commonly known as the Altar of Heaven) , Peking, where the emperor made his annual offerings to heaven at dawn on New Year's Day. Established in 1420, it was radically repaired in 1754. The circular marble structure is built up of three tiers, diminishing in diameter, each surrounded by balustrades and approached by steps from four sides (fig. 79a) . Other altars in this vicinity, such as those of the Temple of the Earth, of the Temple of Agriculture, are merely low, plain terraces without embellishment.

A terrace of little architectural interest but of great interest to the student of ancient astronomy is the so-called Ts'e-ching T'ai, or Star Observing Terrace of Kao-ch'eng Chen, near Teng-feng, Honan Province (fig. 79b) . It is one of the nine terraces built by Kuo Shou-ching in the Yuan dynasty for observing the angle of the sun on the longest and shortest days of the year.

P'ai-lou Gateways

The *p'ai-lou* is a characteristic Chinese feature that, like the Han paired *ch'ueh* piers, dignified an entrance. As an open symbolic gateway it perhaps owed something to Indian influence, having more than a passing resemblance to the gateways at the famous Sanchi stupa of 25 B.C.

The oldest existing *p'ai-lou* is perhaps the one in Lung-hsing Ssu, Cheng-ting, Hopei Province. It is possibly of the Sung dynasty, but the upper part is very much altered by later restorations. Somewhat similar gateways, called *wu-t'ou men*, are mentioned in the *Ying-tsao fa-shih*, and the term is seen often in T'ang dynasty literature. However, the structure did not become very popular until the Ming dynasty.

The most monumental of its kind is the marble *p'ai-lou* (1540) at the head of the avenue of the Ming Tombs, Ch'ang-p'ing, Hopei Province (fig. 80a) . Similar but smaller structures were also erected before some of the tombs of the Ch'ing emperors. Stone *p'ai-lou* in other parts of the country, literally innumerable, can be quite different in character. The five related *p'ai-lou* near Kuang-han, Szechuan Province, are individually typical but a fairly unusual and very impressive sight as a group (fig. 80b) .

Wooden *p'ai-lou* are common in Peking. A street *p'ai-lou* and one in front of the lake at the Summer Palace are quite typical (figs. 80c, d) . Brick masonry arches, embellished with the form of a *p'ai-lou* in glazed terra-cotta, are also a common sight in Peking (fig. 80e) .

Chinese Dynasties and Periods Cited in the Text

中国朝代和各时期与公元年代对照表

Shang-Yin Dynasty 商—殷	ca.1766–ca.1122 B.C.
Chou Dynasty 周	ca.1122–221B.C.
Warring States Period 战国	403–221 B.C.
Ch'in Dynasty 秦	221–206 B.C.
Han Dynasty 汉	206 B.C.–220 A.D.
Eastern Han 东汉	25–220
Six Dynasties 六朝	ca.220–581
Northern Wei Period 北魏	386–534
Northern Ch'i Period 北齐	550–577
Sui Dynasty 隋	581–618
T'ang Dynasty 唐	618–907
Five Dynasties 五代	907–960

图像中国建筑史

（Northern）Sung Dynasty 北宋   960–1127
Liao Dynasty 辽   947–1125（in North China）

（Southern）Sung Dynasty 南宋   1127–1279
Chin Dynasty 金   1115–1234（in North China）

Yuan Dynasty（Mongol）元   1279–1368

Ming Dynasty 明   1368–1644

Ch'ing Dynasty（Manchu）清   1644–1912

Republic 中华民国   1912–1949
Sino-Japanese War 抗日战争   1937–1945

People's Republic 中华人民共和国   1949–

Glossary of Technical Terms
技术术语一览

"accounted heart"	see *chi-hsin*
an 庵	convent
ang 昂	a long slanted lever arm balanced on the *lu-tou*. Its "tail" bears the load of a purlin and is counter-balanced by the eave load at the lower end, in T'ang and Sung construction
ch'a-shou 叉手	truss arms abutting the ridge pole or topmost purlin
che-wu 折屋	the method of "bending the roof"
ch'en-fang t'ou 衬枋头	a small tie beam above the *hua-kung*
chien-chu 建筑	architecture
chi-hsin 计心	a bracket tier or "jump" with complete crossed arms (*kung*) ; literally, "accounted heart." Cf. t'ou-hsin
ch'i 栔	a six-*fen* gap or filler between two *ts'ai*
ch'iao 桥	bridge
chih 楯	transitional disc-shaped member, bronze or stone, between base and foot of column
chien 间	bay

chin-kang pao-tso t'a 金刚宝座塔	five pagodas on a common base, derived from India
ching-chuang 经幢	Dhanari column: freestanding small Buddhist monument in form of column of pagoda
ch'ing-mien ang 琴面昂	"lute-face" *ang* (with pulvinated concave-bevel beak)
chu 柱	column
chu ch'u 柱础	column base or plinth
chu-ju-chu 侏儒柱	small king post; literally, "dwarf post"
chü-che 举折	Sung term: method of determining the pitch and curvature of a roof; literally, "raise/depress"
chü-chia 举架	Ch'ing term: method of determining the pitch and curvature of a roof; literally, "raising the frame"
chü-kao 举高	the height of a curved roof from the lowest to the highest purlin
chuan-chien 攒尖	pyramidal roof
ch'uan 椽	rafter
ch'üeh 阙	paired gate piers (commonly of Han date)
ch'ung-kung 重栱	two arms forming a double tier to support a lintel
Dhanari column	see *ching-chuang*
fang 枋	small beam or lintel
fang-ch'eng ming-lou 方城明楼	memorial shrine on a high square-walled bastion before a tomb (Ming or Ching)
fen 分°	the basic Sung unit for measuring a *ts'ai* (module). One *ts'ai* is 15 *fen* high and ten *fen* wide
feng-shui 风水	geomantic influences; literally, wind and water
fu-chiao lu-tou 附角栌斗	a grouping of *lu-tou* on the plate (*p'u p'ai fang*) to strengthen a corner. Adjacent to the corner *lu-tou*, an additional *lu-tou* is added in each of the two

	directions. One of several techniques for extending complex bracket support around a corner.
fu-tien 庑殿	hip roof
"full ts'ai"	see *tsu ts'ai*
"grasshopper head"	see *ma-cha t'ou*
hsieh-shan 歇山	gable and hip roof
hsuan-shan 悬山	overhanging gable roof
hua-kung 华栱	bracket extending forward and back from the *lu-tou*, at right angles to the wall plane
jen-tzu kung 人字栱	inverted-V bracket
"jump"	see *t'iao*
ke 阁	multi-storied pavilion
kuan 观	Taoist temple
kung 宫	palace
kung 栱	bracket arm: bow-shaped timber, set in a bearing block. It supports a smaller block at each upraised end and often in the center
lan-e 阑额	lintel or architrave
lang 廊	roofed open corridor usually connecting two buildings
liang 梁	beam
ling 檩	purlin
ling-chiao ya-tzu 菱角牙子	chevron-corbelled cornice, on brick pagoda
lou 楼	a building of two or more stones
lu-tou 栌斗	the principal bearing block, lowest in a bracket set
ma-cha t'ou 蚂蚱头	Ch'ing term: shape of beam end; literally, "grasshopper head"
miao 庙	temple（general term）

ming-fu 明栿	exposed beam
ming-pan 皿板	Han term: a square board beneath the *lu-tou*
p'ai-lou 牌楼	freestanding high gateway, joined overhead
pi-tsang 壁藏	wall sutra cabinet
p'i-chu ang 劈竹昂	"split bamboo *ang*"（with straight-bevel beak）
p'ing-tso 平坐	mezzanine story or balcony
pu 步	Ch'ing term; step, the distance from purlin to purlin
p'u-p'ai fang 普拍枋	plate supporting the bracket sets and resting on the lintel, with which it forms a T-shaped cross section
san-fu yun 三福云	cloud-cluster decoration
sha 刹	finial, atop pagoda
shan-men 山门	main gate of a temple
shu-chu 蜀柱	king post
shu-mi-tso 须弥座	high base with decorated mouldings
shu-yao 束腰	narrow dado in a *shu-mi-tso*
shua-t'ou 耍头	head of the beam（protruding）
ssu 寺	Buddhist temple
"stolen heart"	see *t'ou-hsin*
stupa	Indian term: Buddhist burial shrine
t'a 塔	pagoda
t'ai 台	platform or terrace
tan-kung 单栱	one tier of *kung* to support a lintel
t'an 坛	ceremonial terrace or altar
ti-kung 地宫	tomb chamber; literally, "underground palace"
t'i-mu 替木	a narrow member inserted in the *lu-tou* as an auxiliary half *kung* to support the bracket arm. Also sometimes functions as an arm above a bracket set.

t'iao 跳	an upward projection or tier of a bracket set outward or inward; literally, "jump"
tien 殿	monumental hall
t'ien-hua 天花	ceiling
t'ing-tzu 亭子	small pavilion
to-tien 朵殿	smaller detached side halls of a structure on the same axis; literally, "ear halls"
t'o-chiao 托脚	slanting members bracing individual purlins. See fig. 4
tou 斗	bearing block
tou-k'ou 斗口	in an intercolumnar bracket set, the opening in the *lu-tou* to receive the *kung*; literally, "block mouth." Its width is the basic module in Ch'ing dynasty construction. Cf. *ts'ai*
tou-kung 斗栱	bracket set
t'ou-hsin 偷心	a bracket tier or "jump" with *kung* in one direction only; literally, "stolen heart." Cf. chi-hsin
ts'ai 材	a standard-sized timber used for the *kung*; also, module for measurement in the Sung dynasty. Cf. *tou-k'ou*
tsao-ching 藻井	caisson ceiling
ts'ao-fu 草栿	rough beam concealed above the ceiling
tsu-ts'ai 足材	a "full *ts'ai*," namely a standard 15- × -10-*fen ts'ai* increased in height by a 6-*fen ch'i*, making a *tsu ts'ai* 21- × -10-*fen*
tsuan-chien 攒尖	pyramidal roof
wen-fang t'a 文峰塔	good luck tower for examination aspirants
wu-liang tien 无梁殿	beamless hall (barrel-vaulted brick structure)

wu-t'ou men 乌头门	gateway or two or four uprights without a cross beam or roof
yen 檐	eave
ying shan 硬山	flush gable roof
yüeh-liang 月梁	slightly arched beam; literally, "crescent-moon beam"

Guide to Pronunciation

汉字拼音法指南

The two Englishmen who produced the Wade-Giles system for romanizing the sounds of Chinese were naturally thinking of English-language equivalents. Thus a non-Chinese-speaking American or Briton can approximate to some degree Chinese pronunciations by following the W-G spelling. For this we must think of each word as having two parts, initial and fifinal. For certain initial consonants, the aspirate is a signal to breathe out naturally when sounding: for example, p' and t'. Without the aspirate, the signal is: withhold the breath to produce the related but stifled sounds b and d. The initials to which this system is applied are:

p' = p	p = b
t' = t	t = d
k' = k	k = g
ch' = ch	ch = j
tz' = tz	tz = dz

The vowel sounds are the key to the finals. A selected list is given here.

a = ah	*t'a* = (tah) pagoda
	ang = (ahng) slanting lever
ai = i	*t'ai* = (tie) platform
ao = ow	*miao* = (cat-call) temple
e = u, in up	*men* = (mun) gate, door
final e silent	*ke* = (g) high pavilion
i = ee	*pi* = (bee) pen
ie = yeh	*tien* = (dyen) monumental hall
ei = ay	nei = (nay) inside

ui = way   hui = (hway) understand

o = aw   fo = (faw) Buddha

ou = ō   *tou* = (doe) bracket block

u = oo   *kung* = (goong) bracket arm

final u silent   *ssu* = (ss) Buddhist temple

in certain

cases

Selected Bibliography

部分参考书目

Chinese Sources

Shortened Names of Publishers in Peking

Chien-kung: Chung-kuo chien-chu kung-yeh ch'u-pan-she 中国建筑工业出版社（China Building Industry Press）

Ch'ing-hua: Ch'ing-hua ta-hsueh chien-chu hsi 清华大学建筑系（Tsing Hua University, epartment of Architecture）

Wen-wu: Wen-wu ch'u-pan-she 文物出版社（Cultural Relics Publishing House）

YTHS: Chung-kuo ying-tsao hsueh-she 中国营造学社（Society for Research in Chinese Architecture）

Publications

Bulletin, Society for Research in Chinese Architecture. See *Chung-kuo ying-tsao hsueh-she hui-k'an*.

Chang Chung-yi 张仲一, Ts'ao Chien-pin 曹见宾, Fu Kao-chieh 傅高杰, Tu Hsiu-chn 杜修均. *Hui-chou Ming-tai chu-chai* 徽州明代住宅（Ming period houses in Hui-chou ［Anhwei］）. Peking: Chien-kung, 1957.

A useful study of domestic architecture.

Ch'en Ming-ta 陈明达. *Ying-Hsien mu-t'a* 应县木塔（The Ying Hsien Wooden Pagoda）. Peking: Wen-wu, 1980.

Important text, photographs, and drawings by Liang's former student and colleague. English abstract.

——. *Ying-tsao fa-shih ta-mu-tso yen-chiu* 营造法式大木作研究（Research on timber construction in the Sung manual *Building Standards*）. Peking: Wen-wu, 1981.

A continuation and development of Liang's research.

Ch'en Wen-lan 陈文澜, ed. *Chung-kuo chien-chu ying-tsao t'u-chi* 中国建筑营造图集 Chinese architectural structure: Illustrated reference manual）. Peking: Ch'ing-hua （nei-pu）（内部）, 1952. No text; for internal use only.

Chien-chu k'e-hsueh yen-chiu-yuan, Chien-chu li-lun chi li-shih yen-chiu-shih, Chung-kuo chien-chu shih pien-chi wei-yuan-hui 建筑科学研究院建筑理论研究室中国建筑史编辑委员会（Editorial Committee on the History of Chinese Architecture, Architectural Theory and History Section, Institute of Architectural Science）, *Chung-kuo ku-tai chien-chu chien-shih* 中国古代建筑简史（A summary history of ancient Chinese architecture）. Peking: Chien-kung, 1962.

——. *Chung-kuo chin-tai chien-chu chien shih* 中国近代建筑简史（A summary history of modern Chinese architecture）. Peking: Chien-kung, 1962.

Chien-chu kung-ch'eng pu, Chien-chu k'e-hsueh yen-chiu-yuan, Chien-chu li-lun chi li-shih yen-chiu-shih 建筑工程部建筑科学研究院建筑理论及历史研究室（Architectural Theory and History Section, Institute of Architectural Science, Ministry of Architectural Engineering）. *Pei-ching ku chien-chu* 北京古建筑（Ancient architecture in Peking）. Peking: Wen-wu, 1959.

Illustrated with excellent photographs.

Chung-kuo k'e-hsueh yuan T'u-mu chien-chu yen-chiu-so 中国科学院土木建筑研究院（Institute of Engineering and Architecture, Chinese Academy of Sciences）, and Ch'ing-hua ta-hsueh Chien-chu hsi 清华大学建筑系（Department of Architecture, Tsing Hua University）, comp. *Chung-kuo chien-chu* 中国建筑（Chinese architecture）. Peking: Wen-wu, 1957.

Very important text and pictures, supervised by Liang.

Chung-kuo ying-tsao hsueh-she hui-k'an 中国营造学社汇刊（Bulletin, Society for Research in Chinese Architecture）. Peking, 1930–1937, vol.1, no. 1-vol. 6, no. 4; Li-chuang, Szechuan, 1945. vol. 7, nos. 1–2.

Liang Ssu-ch'eng 梁思成. *Chung-kuo i-shu shih: Chien-chu p'ien ch'a-t'u* 中国艺术史：建筑篇插图（History of Chinese arts: Architecture volume, illustrations）. N.p., n.d., ［before 1949］.

——, Chang Jui 张锐. *Tientsin t'e-pieh-shih wu-chih chien-she fang-an* 天津特别市物质建设方案（Construction plan for the Tientsin Special City）. N.p., 1930.

——, Liu Chih-p'ing 刘致平, comp. *Ch'ien-chu she-chi ts'an-k'ao t'u-chi* 建筑设计参考图集（Reference pictures for architectural design）. 10 vols. Peking: YTHS, 1935-1937.

Important reference volumes treating platforms, stone balustrades, shop fronts, brackets, glazed tiles, pillar bases, outer eave patterns, consoles, and caisson ceilings.

——. *Ch'ing-tai ying-tsao tse-li* 清代营造则例（Ch'ing structural regulations）. Peking: YTHS, 1934; 2nd ed., Peking: Chung-kuo chien-chu kung-yeh ch'u pan-she, 1981.

Interpretation of text from field studies.

——, ed. *Ying-tsao suan-li* 营造算例（Calculation rules for Ch'ing architecture）. Peking: YTHS, 1934. The original Ch'ing text of the *Kung-ch'eng tso-fa* tse-*li*, edited and reorganized by Liang.

——. *Ch'ü-fu K'ung-miao chien-chu chi ch'i hsiu-ch'i chi-hua* 曲阜孔庙建筑及其修葺计划（The architecture of Confucius' temple in Ch'ü-fu and a plan for its renovation）. Peking: YTHS, 1935.

——. *Jen-min shou-tu ti shih-cheng chien-she* 人民首都的市政建设（City construction in the People's Capital）. Peking: Chung-hua ch'uan-kuo k'e-hsueh chi-shu p'u-chi hsieh-hui 中华全国科学技术普及协会, 1952.

Lectures on planning for Peking.

——, ed. *Sung ying-tsao fa-shih t'u-chu* 宋营造法式图注（Drawings with annotations of the

rules for structural carpentry of the Sung Dynasty ）. Peking: Ch'ing-hua（nei-pu）
（内部）, 1952.

No text; for internal use only.

——, ed. *Ch'ing-shih ying-tsao tse-li t'u-pan* 清式营造则例图版（Ch'ing structural regulations:
Drawings）. Peking: Ch'ing-hua（nei-pu）（内部）, 1952.

No text; for internal use only.

——, ed. *Chung-kuo chien-chu shih t'u-lu* 中国建筑史图录（Chinese architectural history:
Drawings）. Peking: Ch'ing-hua（nei-pu）（内部）, 1952.

No text; for internal use only.

——. *Tsu-kuo ti chien-chu* 祖国的建筑（The Architecture of the Motherland）. Peking:
Chung hua ch'uan-kuo k'e-hsueh chi-shu p'u-chi hsieh-hui 中华全国科学技术普
及协会, 1954.

A popularization.

——. *Chung-kuo chien-chu shih* 中国建筑史（History of Chinese architecture）. Shanghai:
Shangwu yin-shu kuan 商务印书馆, 1955.

Photocopy of his major general work, handwritten in wartime, ca. 1943. Published
for university textbook use only. Unillustrated.

——. *Ku chien-chu lun-ts'ung* 古建筑论丛（Collected essays on ancient architecture）.
Hong Kong; Shen-chou t'u-shu kung-ssu 神州图书公司, 1975.

Includes his study of Fo-kuang Ssu.

——. *Liang Ssu-ch'eng wen-chi* 梁思成文集（Collected essays of Liang Ssu-ch'eng）. Vol.1.
Peking: Chien-kung, 1982.

First of projected series of six volumes.

Liu Chih-p'ing 刘致平. *Chung-kuo chien-chu ti lei-hsing chih chieh-ko* 中国建筑的类型和
结构（Chinese building types and structure）. Peking: Chien-kung, 1957.

Important work by Liang's student and colleague.

Liu Tun-chen［Liu Tun-tseng］刘敦桢, ed. *P'ai-lou suan-li* 牌楼算例（Rules of calculation
for P'ai-lou）. Peking: YTHS, 1933.

———. *Ho-pei sheng hsi-pu ku chien-chu tiao-ch'a chi-lueh* 河北省西部古建筑调查纪略（Brief report on the survey of ancient architecture in Western Hopei）. Peking: YTHS, 1935.

———. *I Hsien Ch'ing Hsi-ling* 易县清西陵（Western tombs of the Ch'ing emperors in I Hsien, Hopei）. Peking: YTHS, 1935.

———. *Su-chou ku chien-chu tiao-ch'a chi* 苏州古建筑调查记（A report on the survey of ancient architecture in Soochow）. Peking: YTHS, 1936.

———, Liang Ssu-ch'eng 梁思成. *Ch'ing Wen-yuan ke shih-ts'e t'u-shuo* 清文渊阁实测图说（Explanation with drawings of the survey of Wen-yuan Ke of the Ch'ing）. N. p., n.d. [before 1949].

———. *Chung-kuo chu-chai kai-shuo* 中国住宅概说（A brief study of Chinese domestic architecture）. Peking: Chien-kung, 1957.

Also published in French and Japanese editions.

———, ed. *Chung-kuo ku-tai chien-chu shih* 中国古代建筑史（A history of ancient Chinese architecture）. Peking: Chien-kung, 1980.

An important textbook.

———. *Liu Tun-chen wen-chi* 刘敦桢文集（Collected essays of Liu Tun-chen）. Vol.1.Peking: Chien-kung, 1982.

Lu Sheng 卢绳. *Ch'eng-te ku chien-chu* 承德古建筑（Ancient architecture in Chengte）. Peking: Chien-kung, n. d.[ca. 1980].

Yao Ch'eng-tsu 姚承祖, Chang Yung-sen 张镛森. *Ying-tsao fa-yuan* 营造法原（Rules for building）. Peking: Chien-kung n. d.[ca. 1955].

A unique account, written several centuries ago, of construction methods in the Yangtze Valley.

Western-Language Sources

Boerschmann, E. *Chinesische Architektur. 2 vols.* Berlin: Wasmuth, 1925.

Boyd, Andrew. *Chinese Architecture and Town Planning, 1500 B. C. -A. D. 1911*. London: Alec

Tiranti, 1962; Chicago: University of Chicago Press, 1962.

Chinese Academy of Architecture, comp. *Ancient Chinese Architecture*. Peking: China Building Industry Press; Hong Kong: Joint Publishing Co. , 1982.

Recent color photographs of old buildings, many restored.

Demieville, Paul. "Che-yin Song Li Ming-tchong Ying tsao fa che." *Bulletin, Ecole Française d'Extrême Orient 25*(1925) : 213–264.

Masterly review of the 1920 edition of the Sung manual.

Ecke, Gustav. "The Institute for Research in Chinese Architecture. I. A Short Summary of Field Work Carried on from Spring 1932 to Spring 1937." *Monumenta Seric*a 2 (1936–37) : 448–474.

Detailed summary by an Institute member.

——. "Chapter I: Structural Features of the Stone-Built T'ing Pagoda. A Preliminary Study." *Monumenta Serica* 1(1935/1936) : 253–276.

——. "Chapter Ⅱ : Brick Pagodas in the Liao Style." *Monumenta Serica 13*(1948) : 331–365.

Fairbank, Wilma. "The Offering Shrines of Wu Liang Tz'u'" and "A Structural Key to Han Mural Art." *Adventures in Retrieval: Han Murals and Shang Bronze Molds*, pp. 43–86, 89-140. Cambridge, Mass. : Harvard University Press, 1972.

Glahn, Else. "On the Transmission of the Ying-tsao fa-shih." *T'oung Pao* 61 (1975) : 232–265.

——. "Palaces and Paintings in Sung." In *Chinese Painting and the Decorative Style*, ed. M. Medley. London: Percival David Foundation, 1975. pp. 39–51.

——. "Some Chou and Han Architectural Terms." *Bulletin No. 50, The Museum of Far Eastern Antiquities* (Stockholm, 1978) , pp. 105–118.

——. Glahn, Else. "Chinese Building Standards in the 12th Century." *Scientific American*, May 1981, pp. 162–173.

Discussion of the Sung manual by the leading Western expert. (Edited and illustrated without her participation.)

Liang Ssu-ch'eng. "Open Spandrel Bridges of Ancient China. Ⅰ. The An-chi Ch'iao at Chao-chou, Hopei." *Pencil Points*, January 1938, pp. 25–32.

——. "Open Spandrel Bridges of Ancient China. Ⅱ. The Yung-tung Ch'iao at Chao-chou, Hopei." *Pencil Points*, March 1938. pp. 155–160.

——. "China's Oldest Wooden Structure." *Asia Magazine*, July 1941, pp. 387–388.
The first publication on the Fo-kuang Ssu discovery.

——. "Five Early Chinese Pagodas." *Asia Magazine*, August 1941, pp. 450–453.

Needham, Joseph. *Science and Civilization in China*. Vol. 4, part 3: "Civil Engineering and Nautics." Cambridge: Cambridge University Press, 1971. pp. 58–210.
Exhaustive treatment of Chinese building.

Pirazzoli-t'Serstevens, Michele. *Living Architecture: Chinese.* Translated from French. New York: Grosset and Dunlap, 1971; London: Macdonald, 1972.

Sickman, Laurence, and Soper, Alexander. *The Art and Architecture of China.* Harmondsworth: Penguin, 1956. 3rd ed. 1968; paperback ed. 1971, reprinted 1978.
Still a leading source.

Sirén, Osvald. *The Walls and Gates of Peking.* New York: Orientalia,1924.

——. *The Imperial Palaces of Peking.* Paris and Brussels: Van Oest, 1926. 3 vols.

Thilo, Thomas. *Klassische chinesische Baukunst: Strukturprinzipien und soziale Function.* Leipzig: Koehler und Amelang, 1977.

Willetts, William. "Architecture." *Chinese Art*, 2: 653–754. New York: George Braziller, 1958.

生活·读书·新知 三联书店

梁思成作品

·五 种·

《中国建筑史》（通校本）

导言撰写：王军

ISBN：978-7-108-07446-1

定价：128.00元

《图像中国建筑史》（梁从诫 译）

导言撰写：傅熹年

ISBN：978-7-108-07659-5

定价：108.00元

《中国雕塑史讲义》（梁思成 编著）

导言撰写：郑岩

ISBN：978-7-108-07529-1

定价：96.00元

《中国古建筑调查报告》（增补版）

导言撰写：王南

ISBN：978-7-108-07592-5

定价：468.00元 全三册

《〈营造法式〉注释》

导言撰写：王贵祥

ISBN：978-7-108-07727-1

定价：198.00元